SCIENTIFIC DEVELOPMENT IN ISLAMIC CIVILIZATION

About the Author

Dr. Aijaz Ahmad was born to Chaudhary Shafaat Mohammad and Mrs. Hameeda Khatoon on August 16, 1969, at Motipur village in the Sant Kabir Nagar district of Uttar Pradesh. After completing his early education in Motipur, Santha, and Basti, he joined Aligarh Muslim University in 1985 and completed his B.A. (Hons), M.A., and M.Phil. in History. In 1994, he was appointed as a Lecturer in History at YMD College, Nuh, affiliated to Gurugram University, Gurugram. As a teacher candidate, he obtained the Ph.D. degree from Jamia Millia Islamia, New Delhi in Modern Indian History, in 2000. During his service career, he has published 16 research papers in many national and international journals. He has published eleven books. The published books are: 1. Indian Muslims Against British Rule; 2. Sir Syed, Aligarh Muslim University and National Politics; 3. Mewati Rebellion of 1857; 4. Uprising of 1857; 5. History of Mewat; 6. Cultural Heritage of Mewat; 7. Haryana ka Itihaas; 8. Poorv Aupniveshik Bharat Me Vigyaan Tatha Praudyogiki; 9. Triple Talaq; 10. Democracy and Discontent in India; and 11. New Dimensions of Indian Historiography. This book, 'Scientific Development in Islamic Civilization', is his twelfth book.

SCIENTIFIC DEVELOPMENT IN ISLAMIC CIVILIZATION

AIJAZ AHMAD

CONCEPT PUBLISHING COMPANY PVT. LTD.
NEW DELHI - 110 059

ISBN: 978-93-5594-643-0

First Published 2023

Published and Printed by

Concept Publishing Company Pvt. Ltd.
A/15-16, Commercial Block, Mohan Garden
New Delhi-110059 (India)
T: +91-11-41101460
E: publishing@conceptpub.com, *W*: www.conceptpub.com

Cataloging in Publication--Courtesy: D.K. Agencies (P) Ltd. <docinfo@dkagencies.com>

Ahmad, Aijaz, 1969- **author.**
Scientific development in Islamic civilization / Aijaz Ahmad.
pages cm
Includes bibliographical references and index.
ISBN 9789355946430

1. Islam and science--History. 2. Technology--Islamic countries--History. 3. Muslims--Intellectual life. 4. Islamic civilization. I. Title.

LCC BP190.5.S3A36 2023 | DDC 297.265 23

This book is dedicated to my mother Hameeda Khatoon for all she has done in so many ways.

Preface

This book, "Scientific Development in Islamic Civilization", is a collection of scientific thoughts and guidances that has been taught to mankind through Quranic verses and various Prophetic Traditions as a religion, Islam. With the rise of Islam, Muslims have made immense leaps forward in the areas of science and technology. The contribution of Islamic civilization to science and technology from the seventh to the seventeenth century is a noteworthy expansion in human history. Not only did the Muslim Scientists conserved the ancient knowledge, but also transformed it into major new contributions to the basic science and technology. Through their innovations, Muslim scientists significantly contributed to human knowledge in various fields. This contribution was so great that Islamic civilization has been the pioneer of scientific, intellectual, and cultural genius for a long period of time.

In this work, three essential factors have been analysed: (1) the inspiration that the Muslims derived from the teachings of the Quran and Hadith, (2) the origins of the sciences and their effect on the Arabs, (3) the achievements of the Muslim scientists and thinkers in various branches of knowledge. We may refer to these three essential sources of Islamic science one after another. In doing so, one can not ignore the relevance of Islamic sciences to the Mediaeval Europe. Modern science owes a great deal to this early effort of the great Muslim scientists and thinkers.

Islam is a religion that consists of pillars of faith and practices according to the Holy Quran, parts of which were explained by Prophet Mohammad (pbuh) to illustrate a complete understanding of Islam. Believers from one generation to the next have practiced the religion, passing on the faith as well as fostering the development of science and technology. The benefaction of Muslims towards the world's renaissance as well as the development of science and technology has been enormous and worth remembering. Part of this is mentioned in this book, which is a step by step a narration of scientific development by Muslim scientists from the seventh to the seventeenth centuries, which is popularly known as the Golden Age of Muslim civilization.

These scientific thoughts spread throughout the Arab world, which covers a vast geographic area, comprising many different countries in Asia and Africa, and even in some parts of Europe. The contemporary world owes much of its progress in all fields of human intellectual activity, including medicine, to Arabic culture, especially the advancements made during the 'Golden Age' of Arabic-Islamic science. The glorious historic background of the Arabic world permits us to identify the debt that humanity owes to the Golden Age of Arabic science and to evaluate the research contributions made by Arab countries to astronomy, metaphysics, geography, physics, chemistry, biomedical sciences etc.

This book is an attempt to instil the feeling of greatness in the achievements of our past scholars and scientists, neither to be left alone as mere past history nor to be boasted off, but more importantly, to act as a catalyst to motivate us and the future generation to lead the revival of scholastic attitude and scientific advancement in this modern age. This book is, of course, a definition of what Allah says, "I teacheth the man which he knew not." This work especially condemns the Western historians' generalized assertion that Islamic science is Greek or derives from the Greek legacy.

In this work, it will be demonstrated that Islam is the source of all Islamic science and accomplishments and that Greek science was only a tool, a means, just as others were, such as the Hindu and above all, the Chinese legacy. Most of the basic manifestations of Islamic civilization have no Greek antecedent of any sort. It is also an important task and also a responsibility of writing this book that the youth should not forget the commandments of the Quran for the development of knowledge, and scientific research and the important underappreciated scientists from Muslim civilization. The powerful impetus given by the Holy Quran to the study of nature and surrounding realities and the reflection and reasoning, for discovering the signs of Allah transformed the early Muslims into ardent students and researchers.

This work briefly discusses the wide range of approaches and options to utilize and focus knowledge of science, social studies, and the arts by Muslim scholars. This Golden Age of Muslim civilization offers a variety of ways to excite students about science, history, and social studies. This work is a process to understand the scientific approach and inventions of the Islamic world, and relive the one thousand years long adventure of science and technology in Muslim civilization, and provides a positive message for our youth, for the Muslim world, and for humanity. This work displays the history of medicine, astronomy, mathematics, geometry, chemistry, and so on, which still manage to amaze us even today.

After the emergence of Islam, the Muslims managed a domain that stretched from China to Spain for so many centuries. They were the true followers of Islam, and Islam teaches the virtue of the progressive acquisition of new knowledge, as the Quran calls upon man to constantly pray to his Lord for an increase in knowledge. The early Muslims travelled to all corners of the world in their quest for knowledge and took up the study of natural phenomena with religious fervour and commitment, observed nature carefully and undertook extensive experimentation to gain insight and understanding of the nature of things. They gratefully and freely accepted, assimilated, integrated, and advanced the accumulated scientific knowledge, wisdom, and learning of the civilizations that existed at the time. It was in the field of natural sciences that they made outstanding advances and achieved their greatest triumph. This upsurge of scientific activity among the early Muslims is irrefutable proof of the inherently rational and empirical orientation of the Islamic ideology and world-view.

This book sheds new light on the forgotten history and the innovations of Muslim scientists who helped and pave the way for the European Renaissance and the modern world as we know it. It is part of a global effort to increase understanding of how men and women of many different faiths and ethnicities, living under the umbrella of Muslim civilization, made extraordinary advances during this period and how these advances touch every part of society today. This work throws light on an enlightening journey through time to discover how advances in the fields of medicine, optics, mathematics, astronomy, engineering, cartography, and even the rudiments of aviation can trace their roots back to great thinkers from a bygone era.

Within these pages, there is a lost story of scholarship, imagination, and discovery in a time and place where faith, gender, and race were secondary to progress. A man worked beside a woman, Muslim beside a non-Muslim, all in the name of advancing the world for humankind. In fact, science, astronomy, mathematics, algebra, law, history, medicine, pharmacology, optics, agriculture, architecture, theology, music etc. were flourishing in the Islamic world, which would contribute enormously to the European Renaissance.

I have also provided a sketch of the political and economic institutions of the Caliphate and have discussed some of the main stages in the growth of Muslim culture and administration. The history of Islamic philosophy, the development of architecture and the growth of scientific investigation have been analysed in various sections. I have also given a brief account of Muslim Spain, Bagdad, and a few other Muslim cities. These qualities were exemplified in the flowering of Arab genius under the Abbasids, when princes vied with scholars in the translation of Greek classics and the cultivation of science. In the Middle Ages, the fullest development of the Muslim intellect was expressed in the pioneering work of Muslim scientists, historians, geographers, and philosophers.

Fortunately, many Islamic countries and universities are now conducting scientific research and intellectual endeavours. What is called by some people the Islamisation of knowledge, which is now very popular and which goes back to some of the Muslim intellectuals. The treatises written by many Islamic scholars are now being discussed in educational conferences throughout the Islamic world, which is finally going to bear some fruits. However, it will take much more concerted efforts from the most intelligent and gifted members of the Islamic community, who must be well-versed in Western science, Islamic thought, and the Quran's cosmological message. The task is a very daunting and difficult one, but not impossible. I believe that Islam's spiritual and intellectual resources and its holistic vision of scientific enquiry can make a significant contribution to the creation of a new scientific and technological culture that is focused on serving the interests of the whole of humanity.

The purpose of writing this book is also to aware the readers and intellectuals about the scientific temperament and approaches of our forefathers. One of the purposes of writing this work is also to clear the general perception of Muslim intellectuals that we have 700 years of Islamic intellectual history. We can revive the glory of our past and its absurd to think that how can a patient be revived that has been dead for a long time. By study and research of our past glory and of course, the scientific development of the West we can definitely achieve our goal. We can prove our intellectual capability over the propaganda of the West that Muslims were very brilliant, that they did science and things like that. This work is also a process to overcome the very long historical loss of memory.

This work includes the histories of some Muslim nations after the post-colonial era, which came into existence after the downfall of the

Ottoman Empire. It has also been very clearly examined how these nations tried to develop their economies and other infrastructures, but they were brutally suppressed by the Western powers on petty execuses. Perhaps the Western powers never wanted to see another Muslim Golden Age in the future. Now, Muslim countries are either fighting for their existence or blindly following Western nations in terms of policy, economy, and warfare strategy.

I have tried to write this book in a way that will be understandable to educated readers without specialized knowledge of Islam. I have, therefore, avoided assuming much knowledge about Islam and, in particular, the extensive use of Arabic words and names. This book is thus a reminder to my friends and readers that the core intellectual tradition of Islam is deeply rational, though based on revelation. The Islamic view is that religion and science are truly in need of each other. The facts which are included in this book are not so common and most of our Muslim and non-Muslim friends are usually unaware of them and thus misunderstand Islam.

This particular book is entirely the product of an academic study conducted by the author as an academic, researcher, and educator in the past decades. With experience from these roles, I hope that this book will interest all readers. I hope this book will be of benefit to students, scholars, and laypeople, Muslims and non-Muslims alike, who wish to obtain a more holistic understanding of Islam and science for humanity's benefit.

It is also an effort to encourage scientists of all persuasions to cooperate for the betterment of human society and the global environment. It is my fervent hope that the pursuit of science and technology by Muslims will help bring teachers, researcherss and practitioners together to strengthen the Muslim community. Their interactions with non-Muslims will demonstrate that we are enjoined to cooperate in the true spirit of cooperation in pursuing the common good of mankind. Unity is essential for progress on all fronts. Quranic verses prove that unity is the character of Allah's Creation when it is traced back to its primal source. Allah tells us in the Holy Quran that "the heaven and earth were one and we split them apart (Al-Anbiya 21:30)" and "O Mankind! Be obedient to your Lord Who created you from a single living being (*Nafsin Wahidah*) and created from it its mate, and from these two, He created numerous men and women, (Al-Nisa 4:1)".

In completing this work, I have accessed many reputed libraries, archives, and websites, and also interviewed many reputed Islamic scholars. Some of my friends and well-wishers have also boosted and encouraged me to complete this work. I am especially thankful to Chaudhary Zakir Hussain, Mr. S.M. Javed, Prof. Ziaur Rahman, Prof. Waseem Raja, Dr. Mohammad Isa Gauhar etc., for giving every type of help and assistance in completing this book. My heartiest thanks are due to my wife, Imrana Anjum, my kids, Lubna, Iram, Anam, and Sadaf, whose cooperation and sweet smile always encouraged me to do my work continuously.

Aijaz Ahmad

Contents

1
Introduction

This book, "Scientific Development in Islamic Civilization", is a process to make mankind aware of Islamic teachings for the quest of knowledge and scientific research. Islam has its own golden history almost in every sphere of knowledge. Islam enlightens every aspect of human life for the growth, development, and advancement of society. Acquisition of knowledge of humanity, social sciences, as well as pure and applied sciences, has been comprehensively expressed in Islamic education. Almost all the scholars and scientists of the Islamic Golden Age were the great Islamic scholars as well as the great scientists.

Before we discuss Islamic scientific studies and scientists, we must first understand Arabian society, Islam, and the teachings of the Prophet Mohammad (Abu al-Kasim Mohammad son of Abdullah son of Abdul Muttalib son of Hashim). Nevertheless, most of the population of Arabia was mercantilists and pastoralists, who often quarrelled among themselves. It was only during the pilgrimage season to Makkah that fighting was abandoned by common consent. On the whole, the Arabian environment did not encourage the growth of civilized values. It was the emergence of Islam that fostered such a primitive people to emerge from centuries of backwardness to a level of culture. These Arabs not only changed their way of thinking but also their view of the world and their role in it.

Islam means a state of health or of nature. It is derived from 'aslama', meaning "he submitted himself or he entered into peace" or "to give yourself entirely to Allah." Islam is the primordial state of health or nature in which Almighty Allah created man and the universe. Islam is a dynamic process of faith-in-action. A Muslim must always strive to become a better and fuller Muslim by bringing his self-will into accord with Allah's will. It is through submission to Allah's will that peace and the realization of one's destiny can be achieved. Islam is the natural religion of everyone in the universe: "And whoever is in the heavens and the earth makes obeisance to Allah only, willingly and unwillingly, and their shadows too, at morn and eve."

Islam has existed since the time of its first prophet, Adam, but as a systematic religion for the people, it emerged in the seventh century in Makkah, a remote town in Arabia. Here in Makkah, Prophet Mohammad (pbuh) was born to Abdullah and Aminah at 4.30 am on Monday, 22 April 571. Both of his parents belonged to the Quraish tribe. His father died when he was two months old in his mother's womb, while returning from Syria, where he had travelled with one of the trade caravans. Shortly after his birth, the child was taken to the Kaabah and named Mohammad by his grandfather, Abdul Muttalib, a descendent of Abraham through his son Ismael. He lived with his mother for six years before she passed away. After that, he was cared for by his grandfather for two years and was left orphaned again after his grandfather's demise. After that, the person who took over the Prophet's care was his uncle, Abu Talib.

At the age of forty, Prophet Mohammad (pbuh) received his first revelation while in retreat in a cave called Hira, about five kilometres south of the Kaabah: Recite! Recite in the name of thy Sustainer Who created; Created Man from a clot of blood; Recite—for thy Sustainer is the Most Bountiful—Who taught by the pen; taught Man what he knew not (Quran. 96:1-5). This was the beginning of the descent of the Quran as well as the declaration of Prophethood in 610. The revelation continued over the next twenty-three years, culminating a few days before the death of the Prophet in Madinah in June 632, at the age of sixty-three. The Prophet stressed from the start that he was not starting a new religion, but was simply reminding people to revert to the teachings of prophets who had come before him, and that Allah was the God of Abraham, Jesus, and Moses.

The Grand Prophet of Islam is known for his expertise in religion, government, law, politics, military affairs, and trade. Moreover, he was also a great scientist. His mission was not to the Arabs alone, and he was not sent for one age or clime, but for all mankind to the end of the world. The scientific glory of the Arabic nation originated in the Arabian Peninsula in the seventh century, when the preaching of Prophet Mohammad (pbuh) united the Arab tribes and inaugurated the Muslim religion. The Islamic state was formed in 622 when the Prophet moved from Makkah to Madinah.

Within a century of the Prophet's death (632), the new Arabic-Muslim Empire controlled or influenced a large portion of the world, from Southern Europe through North Africa to Central Asia and on to India, and later on to Indonesia and Eastern Europe. In 711, Arab Muslims invaded southern

Spain and a centre of flourishing civilization (Al-Andalus) was created. Another centre emerged in Bagdad from the Abbasids, who ruled part of the Islamic world during a historic period later characterised as the "Golden Age".[1] Generally, tolerant to the faiths, they found in the lands they ruled, the Caliphs oversaw an incredible expansion of knowledge and prosperity. As history shows, Islam as a religion spread more quickly than other religions, and as an Islamic state or empire, it is growing at a similar rate.

It was the inherent truth of the message of Islam, rather than military victories that established Islam across this vast region. It was really miraculous that a few Arab armies, commanding at most a few thousand soldiers, passing through certain selected routes of the ancient world, could bring such a large number of countries and their people into the fold of Islam. By the rapid incorporation of numerous cultures into the fold of Islam, a social revolution was catalyzed. Within those first hundred years, a very large number of micro-transformations took place in the ancient lands of Asia, Africa, and Central Asia to give birth to a new social and cultural mosaic. This multi-cultural, multi-ethnic, and even multi-religious society was to create the conditions for Islamic civilization's seven-century-long scientific enterprise.

This vast social change in the eighth century was the result of mass conversions, migrations, the establishment of new cities and new administrative and fiscal institutions, and a new social contract that Islam established with Jewish and Christian communities. Because of its overarching universal nature, Islam's belief system was able to incorporate diverse social and cultural traditions. The expansion of geographical borders and the subsequent cultural synthesis forced the keenest minds of the times to continuously formulate answers to a wide range of questions arising from.

The rapid geographical expansion also posed new questions related to the practice of faith and that required immediate attention: how to determine the correct *Qiblah* (direction towards the Kaabah in Makkah) for the five obligatory daily prayers from a distant city; how to calculate the correct amount of *Zakah* (obligatory charity on wealth and material goods) on goods; how to apply the principles of inheritance outlined in the Quran to complex situations. This and numerous other issues arising out of a rapidly expanding social, political, and economic landscape produced a fervour of intellectual activity that resulted in the emergence of new branches of knowledge.

The period between the seventh and the fifteenth centuries is considered the 'Golden Age of Islamic Civilization'. During this period, Muslim scientists were successful in their contributions to science and civilization when they maintained the nexus between existential reality and its metaphysical dimension, a theo-centric value order that guided their scientific enquiry, and their quest for knowledge and truth. During this period, there was great emphasis on the pursuit of knowledge. Consequently, there were individuals who lived scholarly and pious lives, such as Ibn Sina, Al-Khwarizmi, and Al-Biruni, who, in addition to excellence in the study of religious texts, also excelled in mathematics, geography, astronomy, physics, chemistry, and medicine. At that time, Islam was not just a set of religious beliefs, but a set of ideas, ethics, and ideals encompassing all aspects of human life. This resulted in the establishment of an Islamic civilization.

The Holy Quran is the main lighthouse of knowledge, and all light and guidance must come from this revealed word of the Almighty. Many Muslim scholars in the Golden Age of Islam studied nature in the context of the Quran. The Quran depicted the relationship between nature and man, and this inspired the Muslim scholars to study natural phenomena, in order to understand Almighty Allah. Islam's contribution to the scientific enterprise was complex and rich, and it spanned over three continents and nearly a millennium of time.

The Islamic view of nature during the Golden Age was for mankind 'to study nature in order to discover Almighty Allah and to use nature for the benefit of mankind'. The Quranic view of nature influenced the development of science in Islamic civilization. Like the Holy Quran, the *Hadith*, or the traditions of the Prophet, is also one of the primary sources of Islam. The Holy Quran and *Hadith* emerged in Islamic civilization prior to any other branch of knowledge, and they influenced all other fields, including the natural sciences, in numerous direct and indirect ways. Thus, science, as a systematic study of nature and as it developed in Islamic civilization, could not treat nature and its study as an entity separate from Islam.

It is impossible in Islam to conceive of nature as an independent, self-subsisting entity. Likewise, science, as an organised enterprise that studies and explores the natural world, cannot be conceived as a separate entity which has to be somehow externally related to Islam. Thus, by the time the study of nature appeared in Islamic civilization as an organised and recognisable enterprise, the religious sciences had already been firmly established; this sequence affected the framework used to explore nature. So the Arabs studied historiography and geography, and they studied the antiquities, archaeology, traditions, and social conditions, not merely of Muslim, but also of non-Muslim countries. The study of history led to the study of archaeology, geography, ethnography, and other sciences.

As trade flourished, so did knowledge and new ideas. Scholars worked to translate the writings of ancient thinkers like Brahmagupta, Aristotle, Euclid, Ptolemy, and Hippocrates into Arabic to allow debate and development of mathematics, astronomy, chemistry, medicine, and engineering. It was a Golden Age of thought, development, and wealth creation. The Muslims preserved, explained, and developed the conception of democracy, they built up the foundations of culture; they preferred the tangible and concrete to the visionary and abstract, the definite to the indefinite, the sensuous to the ideal. The Golden Age of the Abbasids and the Umayyads of Spain had yielded to the world a second crop of flowers and fruitage similar to that which had been produced in Greece and Rome.

The School of Astronomy at Bagdad in the tenth and eleventh centuries carried research of the highest order and greatly enlarged the knowledge of heavenly phenomena. The Astronomical School of Cairo followed in the footsteps of the Bagdad school, and some of them, such as Ibn Yunus, a pupil of Abul Wafa, made new discoveries, evolved new methods, constructed new instruments, and prepared astronomical tables of great importance. Arab astronomers in Spain also carried on the tradition of the Bagdad school.[2]

The Middle Ages came to know of alchemy through the Arabs. The Arabs greatly admired pharmaceutical knowledge, and an important work by the Persian, Abu Mansur Muwaffak, at the end of the tenth century, remained authoritative for centuries. On the whole, it must be admitted that the Arabs made many important discoveries in chemistry. Their chief merit lies in developing the scientific method and collecting a large amount of data. The same remarks apply to medicine. They were to some extent restricted in the field of medicine as religion forbade them to dissect dead bodies, and one of the most important sources of investigation was thereby closed. They based their system entirely on Galen (Jalinus) and it is supreme in many Muslim countries.[3]

In the fifteenth century, Muslim sectarian conflict gave a death blow to scientific learning and superseded fundamentalism. The deadly animosity between Orthodox Muslim scholars and Mutazilites and Rationalists, the emergence of mysticism, the negligence of the rulers etc., led to the downfall of the science and research. Theologians were too preoccupied with fighting rival creeds, with Sunni divines wrangling with Shias and Shia Mujtahids fighting Sunni divines, to find time to purge Islam of superstitions, allowing it to resume its tradition of being a progressive and dynamic religion. The principles of toleration and fraternity which Islam had inculcated were completely forgotten by ambitious monarchs who used religion as a cloak for imperialistic designs.

The Muslim world in the post-Mongol era gave rise to three powerful empires: the Safavi 1135-1722, the Indian Timuri 1274-1857, and the Ottoman 1343-1924. These empires extended over a vast area and were extremely rich in resources. They patronised the arts and sciences and vied against each other to attract the best minds of the time. In addition, several other smaller states and dynasties supported science. Yet, none of these empires was able to compete with the developments in science that were taking place after the fifteenth century in Europe. When they did realise the enormous military and economic benefits that Europeans had reaped with their science and technology, it was already too late; better-trained European armies, equipped with superior weapons, were already knocking at their doors.[4]

Islam patronised and fostered the Greek scientific heritage in the fields of medicine, astronomy, mathematics, physics, chemistry, and philosophy. Islam continued to add new scientific achievements, which bear witness that Muslims were deeply and seriously interested in scientific research. In modern times, when Europeans began to study the sciences and philosophy after the renaissance, they translated the works of Muslim scientists and studied them very deeply. Here we can say that, firstly, Greek science was taken over by Muslims, improved and Islamized; then, after the Crusade and Renaissance, Islamic science was taken over by the Latin West and later on the European West.[5]

While the progress of scientific knowledge in Europe languished during the Dark Ages, science flourished in the Golden Age of Islam. The Renaissance that subsequently occurred in Europe might not have taken place without the contribution of Muslim science in the preceding period. The Europeans not only adopted their knowledge, but they even changed the names of the Muslim philosophers and scientists. Al-Ghizzali, for instance, became Algazel; Abu Ma'shar became Albumasar; Ibn Tufail became Abubacer; Al-Bitruji became Alpetragius; Al-Farghani became Alfraganus; Razi became Rhazes; Abu'l-Qasim al-Zahravi became Albucasis; and Ibn Sina became Avicenna. We can see that Arabic names were either carelessly Latinized or were simply distorted because of a lack of proper linguistic skills. This continues in many works that do not use accurate transliteration schemes.

I have divided this book into nine chapters, and the first chapter is introductory and provides a brief survey of the whole book. In the second chapter, I discussed the Commandments of Almighty Allah in the Holy Quran about the creation of the universe—Heaven and Earth—including the living beings. In the third chapter, the Prophetic traditions about teaching, learning, and the development of science have been widely discussed, with special reference to Tibbe Nabvi. The educational and scientific development under the Caliphs of the Ummayyad, Abbasid, and Othmania dynasties is thoroughly discussed in the fourth chapter. In the fifth chapter, the debate over science and rationalism in Islamic theory and practice has been fully covered. Their differences with the Imamates and mysticism have also been covered. The role of Muslim scholars in the development of science and technology under various Islamic rulers of different dynasties has been covered in the sixth chapter. The seventh chapter discusses trade and industry in the Mediaeval Islamic world. The eighth chapter sheds light on science and technology in the modern Muslim world. The last or ninth chapter covers the whole period of the post-colonial Muslim world. In this chapter, the scientific developments in the Muslim countries as well as their devastation by the Western world have been fully analysed.

After summing up the whole work, it could be analysed that the Muslim sciences were promoted by their religious beliefs, helped by Greek philosophy and scientific thoughts, and paved the way to the European or Western Renaissance. This Renaissance developed science and technology so quickly and was never seen back until the contemporary and current period. During these periods, the Western world brutally suppressed the Muslim powers that they felt were their competitors. Actually, the Western powers learned their lesson from history and never wanted another Golden Age of Muslim civilization to occur.

NOTES

1. Matthew E. Falagas, Effie A. Zarkadoulia, and George Samonis, "Arab Science in the Golden Age (750-1258) and Today", *The FASEB Journal*, Vol. 20, August 2006, p. 1581.
2. Bhargava, K.D. (1961). *A Survey of Islamic Culture and Institutions*, Allahabad, pp. 235-36.
3. *Ibid.*, pp. 237-38.
4. Iqbal, Muzaffar (2009). *The Making of Islamic Science*, Kuala Lumpur, p. 154.
5. The earliest Greek scientists were Thales, Anaximander, and Anaximenes. Thales (c. 600 BCE) believed that water was the essence of natural phenomena. To him, matter came in three forms: mist, water and earth. He also thought that the stars were made of water. His pupil, Anaximander, (c. 545 BCE), was believed to have written the earliest book on science, which claimed that life originated in the sea. Anaximenes (c. 500 BCE), a pupil of the former, thought that air was the essence of the universe and that a rainbow was a natural phenomenon and not a divine sign. Aristotle is

generally thought to be the father of the life sciences. He studied 540 plants and classified plants and animals. He also wrote on embryology. Aristotle believed that the earth was the centre of the universe. Aristotle thought that motion was created by an object trying to reach its natural place. The greatest Greek contribution to medicine was made by Hippocrates of Cos, an author of many books, whose Hippocratic Oath is still used as a code of ethics by the medical profession. He freed medicine from superstition and religion. Greek medicine also spread to Rome, where the physician Galen, through his teaching and prolific writings, popularized. The Greek scientist Empedocles formulated the idea of the elements (air, water, earth, and fire), which were adopted by Plato and Aristotle. To Plato, geometry was the most suitable method of thinking about nature. Euclid of Alexandria, the author of the Elements, was the most influential Greek geometrician. The Greeks made important contributions to mathematics, which is a science based entirely on reason, with no need for observation or experiment. Pythagoras (fifth-century BCE) regarded mathematics as the most important branch of science. Diophantus was regarded by some as the Greek founder of 'algebra' (although the term itself had an Arabic origin). Archimedes founded mathematical physics and discovered the laws of hydrostatics. He also invented the Archimedian Screw, a device designed to raise water for irrigation. The Alexandrian engineer Hero was credited with the invention of a series of automata. Ctesibius is thought to have been the founder of the Alexandrian School of Engineering. The Greeks also built a water-carrying tunnel through a mountain. Philon was credited with some technical achievements, including a force pump and a mechanically driven water clock. Ptolemy, a great Greek astronomer from Alexandria, wrote the *Almagest*, which described the planetary motion and placed the Earth as the centre of the Universe, with the Sun and the Moon revolving around it. In 270 BCE, Aristarchus of Samos challenged Aristotle's geocentric idea, asserting that the Sun was the centre of the solar system. He also emphasised that all other planets revolve around the Sun (Beg, Muhammad Abdul Jabbar (ND). *The Origins of Islamic Science*, Leicestershire, p. 16).

2
Holy Quran and Science

The Holy Quran is words of Allah, literally meaning "Readable" or "*Al-Zikr*", revealed to the Prophet[1] Mohammad (pbuh)[2] for all mankind. The Quran is not a book of science but a book of signs. However, there are more than six thousand "signs" in the Quran that deal with science. One of the most remarkable things about the Quran is that it contains many verses that correctly describe natural phenomena in various disciplines, such as human embryology, meteorology, astronomy, geology, and oceanography. Many of the processes and functions mentioned in the Quran have been discovered only recently.

At the age of forty, Mohammad (pbuh) received his first revelation while in retreat in a cave called Hira, about three kilometres south of the Kaabah: Recite! Recite in the name of Thy Sustainer Who created; He created Man from a clot of blood; Recite-for Thy Sustainer is the Most Bountiful—Who taught by pen; He taught Man what he did not know (Quran 96:1-5).[3] From that time, the Quran, the collective revelations, verbally revealed by Allah to Prophet Mohammad (pbuh) over a period of approximately 23 years of Islamic propagation. Its content is concerned with basic Islamic beliefs. Narratives of the early prophets, ethical and legal subjects, and historical events also appear in the Quran.

This was the beginning of the descent of the Quran. Hearing the verses of Allah from Prophet Mohammad (pbuh), his followers memorized, recorded or wrote them down on different materials such as papers, leather, wood plates, rock plates, date branches, bones, earth and many other substances. To combine the verses of the Quran, Prophet Mohammad (pbuh) decided to impose *Surahh* or chapter and *Ayat* or verses, ordering his followers to compile scattered verses from reciters as well as those written on many different materials into one place. As a result, the Quran was gathered into distinct chapters and verses. It was the preface to the development of the barbarous Arab tribes into a highly civilised state, which managed to create the foundations of modern science and technology.

The revelation continued over the next twenty-three years, culminating a few days before the death of the Prophet in Madinah in June 632, at the age of sixty-three. After Prophet Mohammad's (pbuh) death, the Caliphs or supreme leaders of the Muslim state, consulted other leading figures about the collection of the verses of the Quran, which were initially recorded on many different materials such as palm leaves, leather, wood plates, papers, rocks, and others. The third Caliph Othman Ibn Affan, combined the verses of the Quran, which were written at many different places in many different forms of writing and pronunciation during Caliph Umar's reign. His efforts resulted in the final form of the Holy Book Quran.

It may be notable that the Islamic enlightened route started with "Read", followed by five Quranic verses consulting about science. A statistic for the number of letters in the word "science" and its derivations in the Noble Quran, it was repeated 779 times, that is an average of 7 times a chapter. It is placed the second after the word "Allah". This, in fact, replicates the importance of knowledge in the Islamic civilization. For the promotion of the knowledge, Quran (20:114) says, "O Lord! Increase me in knowledge."

Prophet as a Messenger of Allah spread and promoted religion among the people. Unlike other religions, Islam claims that it is the final form and, hence, there will be no need for any more prophets to rediscover the nature of eternal truth. There are three essential elements in Islam: (i) belief in one God, the Creator, Originator, and Sustainer of all things; (ii) belief in Prophet Mohammad (pbuh) as the messenger of Almighty Allah and his messages; and (iii) belief in the eventual end of all things, ushering in another kind of life in the Hereafter. In other words, it may be said that belief in *Tawhid* (Oneness of God), *Risalah* (Prophethood), and *Maad* (Return). These are presented by the Quran in various ways and are supported through numerous demonstrational proofs.

Islam, as developed during Prophet Mohammad's (pbuh) time, is a religion of emancipated individuality, and in it human personality attains its fullest development. Islam's emphasis on the unity of Allah formed the basis of its scientific outlook. Science is a civilizational activity to study and explore the natural world; it fulfils the needs of a given civilization by providing reliable and verifiable knowledge about the physical world. The enterprise of science cannot be an autonomous undertaking; it is always part of a civilization. In traditional civilizations, science was always part of a hierarchy of knowledge that paid attention to the physical world in due proportion. It is pursued by men and women whose understanding of the physical world they explore is directly related to their belief system and worldview. So there are a lot of similarities between religion, Islam, and scientific knowledge. In fact, Quran, in many of its verses, commands the people to seek knowledge and scientific exploration.

Quran and Science

The Holy Quran was revealed to Prophet Mohammad (pbuh) within twenty-three years (610-632). It was for mankind in the whole world and established the law of relations between man and Allah. The Quran is the main lighthouse of knowledge, and all light and guidance must come from this revealed words of the Almighty. The Holy Quran guides mankind to the right path of livelihood, acquires knowledge, and scientifically studied the resources of the heavens (universe) and earth. About 750 of the verses of the Quran in different chapters are related to science and technology, astronomy, geology, sanitation, medication, food and nutrition, while the basic content is about history, law, government, faith and other religious performances. Four attributes of Almighty Allah mentioned by the Quran are especially relevant to knowledge and research: *Al-Khaliq* (the Creator), *Al-Bari* (the Maker), *Al-Musawwir* (the Shaper), and *Al-Baat* (the Originator).

T.V.N. Persaud, a professor of several medical fields at University of Manitoba, Canada, with a number of academic works as well as medical prizes, made mention of Mohammad (pbuh) in a book he authored, *This is the Truth*, edited by Abdullah M. Al-Rehaili and published in 1998. He wrote that in regard to sciences, especially medicine, the Quran has many verses that revolve around knowledge of modern day medicine. He concluded that the book could not have come from anywhere else but from Allah through an illiterate, ordinary man named Mohammad (pbuh). There is no evidence that this man learned anything about science in his lifetime. About 12 per cent of the Quran has been acknowledged by both Muslim and non-Muslim scholars to be related to science, mostly medical science. Dr. Persaud confirmed that none of the verses contradicts modern day findings. Besides, much research has been conducted to compare the contents of the Quran with several fields of modern knowledge, including science, medical science, engineering, astronomy, chemistry, physics, and so forth.[4]

The universe was created with a purpose as a necessary environment in which man might fulfil the divine trust which he had accepted and started learning. The Holy Quran, revealed to Prophet Mohammad (pbuh) by Almighty Allah, better explains the purpose, aims and objectives of the Heavens and Earth, and also the creation of human being. Allah says that the universe is alive, singing the praises of God. In Surahh al-Nur, 24:41, Allah states, "Do you not see that it is Allah Whose praises all beings in the Heavens and on Earth do celebrate, (including) the birds with their wings outspread? Each one knows its own (mode of) prayer and praise".[5] Besides the praise of Allah, all creatures have mind and brain for their progress and advancement, which he uses in the guidance of the Holy Quran and for mankind, that is science and technology. The Holy Quran and hence Islam raise the human mind to the highest rank to discover the hidden knowledge of the Heavens and Earth.

Man regards this world as a productive farm or field for doing good work for the sake of improving and benefiting earthly life as well as ensuring happiness in the hereafter. Its inimitable elocution and teachings are conspicuous and convincing at all stages of civilization. Since the Holy Quran's teachings are not restricted to a specific age of time or people, the treasures of its unlimited inexhaustible dimensions are continuously realised as time marches on, in eternal reality, vitality and freshness. The quest for knowledge and study quickly developed in early Muslims with the inspiration of the Holy Quran and the Traditions of the Prophet Mohammad (pbuh).

Already during the life of the Prophet, the sciences of the Quran (*Ulum al-Quran*) had emerged as a differentiated branch of learning. Shortly after his death, focused efforts were made to preserve, annotate, and verify the *Hadith* (the deeds, sayings, ideas, permissions, and disapprovals of the Prophet). The Holy Quran and *Hadith* are the two most important and fundamental sources for understanding Islam and its philosophy. During the Golden Age, the Islamic view of knowledge of nature or science had its roots in the Quran, the very word of Allah and the basis of Islam. Muslim scholars at that time were inspired to study nature in the context of the Quran.

Most of the passages from the Quran illustrate the relationship between nature and man and how this relationship inspired Muslim scholars to study natural phenomena, in order to understand Almighty Allah. These verses also show the way the Quran presents the whole universe: "We created not the heavens, the earth, and all between them, merely in (idle) sport. We created them not except for just ends: But most of them do not understand (Quran 2: 38-39)".[6] Thus, it was concluded that Allah created the world and placed man in it as trustee, to benefit from it, to use it wisely and to understand his purpose in the universe. Thus mankind was inspired to study, understand, and mould the natural forces for its own purposes. The point to note is the general empirical attitude of the Quran, which engendered in its followers a feeling of reverence and thus made them the founders of an enlightened society.

After the decline of Greece and Rome, these were the Arabs who kept the torch of learning burning brightly in the Middle Ages. The Islamic scientists acquired Greek and Roman knowledge and developed it through Quranic research and knowledge of *Hadith*. A modern Islamic scientist, Nasr, says that Islamic science came into being from a wedding between the spirit that issued from the Quranic revelation and the existing sciences of various civilizations that Islam inherited and which it transmuted through its spiritual power into a new substance, at once different from and continuous with what had existed before it. The international and cosmopolitan nature of Islamic civilization, derived from the universal character of the Islamic revelation and reflected in the geographical spread of the Islamic world, enabled it to create the first science of a truly international nature in human history.[7]

Scientific knowledge fully developed and flourished in the Islamic world from the eighth to the eighteenth century. It was a time of incredible growth and achievement in Muslim thought and understanding. During this period, Muslims developed a great thirst for learning, a craving the likes of which history had never known before. Islamic civilization had reached its zenith and Muslims had become world leaders in philosophical and scientific thought.

The Holy Quran comprises the arguments and evidence on which we also rely in this age of science. By science, we mean a chronological and systematic study of all branches of knowledge. It includes studies concerned with the universe and allied subjects under which fall the modern sciences, such as biology, chemistry, physics, astronomy, meteorology, and geology. The Holy Quran raises the prestige and value of these sciences and encourages people to study them for the general good. A great number of verses not only speak highly of science and knowledge but also urge us to pursue science and to discover the secrets embedded in the build-up of the universe.

After all, one can realise that science is the message of Islam, and the Holy Quran describes with scientific precision matters pertaining to the universe. It also urges Muslims to take an interest in science. In this manner, many famous Muslim scientists arose during the so-called middle ages or even dark ages.

1. *Knowledge and Learning*

Knowledge is *Ilm* in Arabic, a word that frequently occurs in the Holy Quran. The Holy Quran and of course, Islam speak too much about seeking knowledge and learning about the nature, Heavens and Earth. Knowledge is considered meritorious and the Quran tells us that those who know and those who do not know are not equal (Quran 39:9).[8] As it is well-known that science is the systematic knowledge of nature and, undoubtedly, a product of the human mind, the Holy Quran encourages the people to acquire knowledge and science. By adopting scientific methods, we are able to uncover the hidden signs of the Creator and discover His attributes in the material universe, an outstanding step forward towards becoming believers.

Despite numerous observations, scientists are still unable to declare the existence of life in the universe beyond our earth and solar system. A great number of verses in the Holy Quran express the same meaning that we are not alone in the universe. However, logically speaking, one can argue that wherever suitable environments prevail on other planets in the universe, life can nourish and flourish with time. Naturally, such planets may be found only in solar systems other than ours. These verses also mean that beings similar to man exist on some other planets and should be able to communicate with each other by any means when Allah wishes. Of course, the verses quoted above need further discussion and explanation.

It is well known that religions other than Islam were a barrier to scientific progress during the Middle Ages, and the accusation and trial of Galileo (1615) is an example. In Islam, however, there was the reverse matter. The Holy Quran, not only invites people to cultivate science, but also includes numerous verses which refer to several natural phenomena, establish the principles of basic science and favour the increase of knowledge. Regarding this, the Holy Quran mentions in Surah Yunus, "Say, Behold all that is in the heavens and the earth ... (Quran, 10:101)".[9] As for the scientific research and observation, the Holy Quran says in Surah Al-Baqara, "Say, Bring forth your proof, provided you are right.... (Quran, 1:111)"[10]

The Holy Quran itself bears testimony to the supreme value of learning and science. Commenting on the Suraht-ul-Alaq, "Allah taught human beings that which they did not know, and this testifies to the greatness of His beneficence, for He has given to His servants' knowledge of that which they did not know. And He has brought them out of the darkness of ignorance to the light of knowledge, and made them aware of the inestimable blessings of the knowledge of writing, for great benefits accrue there from which Allah alone compasseth; and without the knowledge of writing no other knowledge (*Uloom*) could be comprehended, nor the sciences placed within bounds, nor the history of the ancients be acquired and their sayings be recorded, nor the revealed books be written; and if that knowledge did not exist, the affairs of religion and the world, undoubtedly could not be regulated (Quran, 96:5)."[11]

Allah chose mankind as the best creation on earth in terms of acquiring knowledge and scientific observations, but Allah also mentions some shortcomings in human senses and knowledge. The Holy Quran in Surah Al-Ahzab says, "Verily, We proposed to the Heavens and to the Earth and to the mountains to bear the responsibility of knowledge, but they refused the burden, and they feared to receive it. Man undertook to bear it, but has proved to be unjust and senseless (Quran, 33:72)."[12]

Evidently, however, the Holy Quran mentions the various phenomena of nature and urges mankind to contemplate and study them. The Holy Quran is the eternal message of Allah to mankind for guidance. Allah not only bestowed knowledge on mankind, but he also stressed the importance of making them eager for knowledge, true knowledge of science and technology. It not only means that our studies in science should not be merely for the sake of understanding the material universe, but also that it means of recognising Allah, the Creator of the universe, and thus deepening our faith as believers. The Holy Quran in Surah Taha says, "... And say (asking Allah), O my Master, I beg to increase my knowledge (Quran, 20:114)".[13] Allah again says in another Surah Al-Mujadila, "Allah raises up (to ranks) those of you who believe and who have been granted knowledge (Quran, 58:11)".[14]

It is also worth mentioning that Allah has not fixed a limit of knowledge or scientific progress but has shown in the verses of the Holy Quran a vast field of knowledge which is being gained by the human race through observations and discoveries. Man is continuously gaining knowledge throughout the progress of the civilizations and the availability of infrastructure and observations. The Holy Quran says in Surah Baqara, "And He taught (the race of) Adam, the nature of all things (Quran, 1:31)".[15]

Most of the Quranic verses dealing with scientific issues constitute as a whole the aspects of the Quran's miraculousness that are manifested in this age in which people and communities believe mainly in science, and the strength of nations is gauged by the culture they achieve, the knowledge they acquire, and the inventions they accomplish. Indeed, that is one of the magnificent qualities of the Holy Quran, the Book whose miraculousness is neither confined to a particular age nor ends at a fixed time. Now the way of ascension to Almighty Allah is itself the way of sound knowledge and true science as revealed by the study of mathematics, astronomy, physics, meteorology etc. as stated before in Islam and Science, the very first verses in the Quran heralded the emergence of this Age of Science and urged the search for knowledge.

The most sublime reference regarding knowledge and science and the most outstanding reality in this connection is that the very first verses of the Holy Quran as revealed by Allah gave incentive to gain knowledge and to glorify it. The very beginning of the Revelation was actually a signal for the dawn of the age of science and a forerunner of giving knowledge its due dignity. Thus, the first verse of the Holy Quran says in Surah, Al-Alaq, "Read in the name of Allah the Creator, who created man from a clot. Read and your Master is most generous, who taught by the pen, taught man what he knew not (Quran, 96:1-5)".[16]

The Quran explains in succession, at different stages of the revelation, the meaning of science and education, and points out the importance of studying the universe, its matter, energy, and living creatures. These are the sciences by which man gains faith and fear of Allah, in addition to gaining power. It is a fact that the scientific renaissance was due to gathering correct observations and performing sound experiments. Physical laws are nothing but interpretations of phenomena in the material world. They represent the greatest possible amount of authenticity and precision. In the light of growing accuracy in observation, scientists introduce, from time to time, amendments or generalisations to some of the physical laws so as to bring them closer to what we may call fact, or to make them more useful.

The Holy Quran not only directs man to learn, but it also suggests learning about the various branches of knowledge and their specifications that Almighty Allah has created in Heaven and Earth. In many places, the Quran says something about the inherent relationship among Allah, humanity, and nature. In a chapter of the the Quran entitled Al-Rahman, a vast range of phenomena of the natural world—the sun and the moon, rivers, oceans, fruits, cattle etc.—are mentioned as being so many divine favours and blessings. The Quran then asks, in a powerful refrain occurring thirty-three times in this chapter, which of the favours of your Sustainer will you then deny? (Quran, 55: 1-78).[17]

Many of the Quranic verses show that in Islam it is not wise for people to have mere fantasy as the basis of their religious ideologies or scientific theories. Any conclusion which has no support from experience or clear evidence will not do. They will be like those who deduce the properties of matter or explain natural phenomena without making observations or carrying out experiments, or those who inhibit beliefs without subjecting them to text to determine what is right and wrong.

By thoroughly reading the Quranic verses, it becomes evident that science is the message of Islam, and the Quran describes it with scientific precision and accuracy. The Quranic view of nature influenced the development of science in Islamic civilization. The Quranic verses urged Muslims and encouraged them to take an interest in science during the so-called middle-ages or dark ages. Thus, science, as a systematic study of nature and as it developed in Islamic civilization, could not treat nature and its study as an entity separate from Islam.

2. *Astronomy or Cosmology*

The Holy Quran mentions that Allah created the universe and earth, day and night, and emphasised the people's duty to understand their minds, study, and gain knowledge. The Holy Quran played a key role in establishing a nexus between astronomy and Islam. A verse of the Quran established the lunar year as consisting of twelve months, four of which were specified as sacred (Quran, 9:32); another verse (Quran, 2:149-50) changed the direction of the *Qiblah* from Jerusalem towards the Kaabah in Makkah, requiring Muslims to face this direction for the ritual prayers and certain other acts of worship.

The Quranic injunction to establish *Salah*, the ritual prayers, at specific times, visibility of the moon, also caused the development of a special branch of religious astronomy. It is useful for finding the hours of the day and night, their lengths, and the ways in which they vary. It is also useful in finding the times of prayer and in determining the direction in which one should pray. This science is also concerned with shadow lengths and the altitudes of celestial bodies, and with the orientation of one city from another.

(i) *Heaven or Universe*: Although the Quran is not a text book of science, its greatness lies in the guidance and the framework it provides for an integrated and holistic approach to life and for acknowledging the power, greatness, and kindness of the Creator. However, Quranic cosmological data, the sayings of the Prophet and reflections on these two primary sources by the Companions of the Prophet gave birth to an Islamic cosmological or astronomical tradition. This was later supplemented by scientific observations.

The twenty-ninth verse of the second chapter of the Quran mentions the creation of the seven heavens in a succinct manner: He (Allah) is the

One Who created for you all that is on Earth; then He paid attention to the sky and fashioned it as seven perfect (heavens); He has full knowledge of everything (Quran, 2:29).[18] Commenting on this verse, Ibn Abbas is reported to have said that "Almighty Allah's Throne was on water when nothing other than water had been created yet. When He decided to bring creation into being, He brought out smoke (or, steam) from the water, which raised upwards, and from this He made the skies. Then the water dried up and He made one earth from it; then He fashioned seven earths from this one in two days .... likewise, he fashioned seven heavens from the one He had made from the smoke".[19]

The Holy Quran offers a more specific description of astronomical knowledge: Allah created the Heavens and the Earth, as well as everything in between, in six days (Quran, 7:54-56; 25:59);[20] He created the earth in two days (Quran, 41:9);[21] and placed therein firm mountains (towering) above its surface and bestowed blessings on it, and equitably apportioned its means of sustenance in four days (Quran, 41:10);[22] He turned towards the heaven, which was (yet but) smoke; and He said to it and the Earth—'come both of you, willingly or unwillingly' to which both responded, 'we do come in obedience.' And He then decreed that it (the smoke) becomes seven heavens in two days, and He assigned a function to each heaven (Quran, 41:12);[23] He created the seven heavens (Quran, 2:29);[24] one upon another (Quran 67:3);[25] through them flows down from on high, His Will (Quran, 65:12).[26] Allah has adorned the sky with stars (Quran, 67:5)[27] and is the One who has set in motion all the stars and planets so that humanity may be guided by their positions in its travels (Quran, 6:97);[28] He is the One Who covers the day with the night and the night with the day (Quran, 7:54; 39:5).[29] He has created the day and the night (Quran, 21:33),[30] the sun and the moon, each revolving in its precise orbit (Quran 39:5).[31]

Although the Quran does not explain how or when the cosmos was created, it does invite its readers to study the physical world. In fact, this Quranic invitation to observe the workings of the cosmos is repeated with such insistence that it has now become commonplace to use these verses of the Quran to support the claim that the cultivation of modern science is a religious obligation for Muslims—a duty prescribed by none other than the Quran itself.

The Quran says in Al-Imran, "Verily in the creation of the Heavens (universe) and the Earth, and in the variation of night and day, are signs for people of understanding minds (Quran, 3:190)".[32] In Surah Al-Jathiya, Allah again says, "Assuredly, in the Heavens and the Earth are signs for the believers; and in the creation of yourselves and the living beings He disperses on earth are signs for the firm in faith; and in the variation of night and day and the sustenance which Allah sends down from the Heavens, whereby He gives life to the earth after its death, and in sending winds in their ways are signs for a people of discernment (Quran, 45:2-5)".[33]

For a long time, the origin of the universe and the evolution of life on earth has been a hotly debated or rather unsolved mystery. In this regard, the Holy Quran in Surah Al-Anbiya says, "Do not the infidels see that the Heavens and the Earth were joined together as one unit before we split them apart and that we made of water every living thing? (Quran, 21:30)"[34]

The Holy Quran declares in more than 35 verses that the heavens, away from the solar system, are inhabited by wise beings. Most likely, they resemble human beings on earth in many ways. The Holy Quran in Surah Al-Israa says, "... And your Master best knows those who are in the Heavens and the Earth... (Quran, 17:55)".[35] Again, in Surah Al-Nur, Allah says, "Seest thou not that Allah is He Whom do glorify all those who are in the Heavens and the Earth (Quran, 24:41)".[36] In Al-Shura Allah says, "And of His signs is the creation of the Heavens and the Earth and what He has spread forth in both of them of living beings; and He is all-able to gather them together when He wills (Quran, 42:29)".[37]

A great number of verses in the Holy Quran express the same meaning that we are not alone in the universe. However, logically speaking, one can argue that wherever suitable environments prevail on other planets in the universe, life can nourish and flourish with time. Naturally, such planets may be found only in solar systems other than ours. These verses also mean that beings similar to man exist on some other planets and should be able to communicate with each other by any means when Allah wishes. Of course, the verses quoted above need further discussion and explanation.

It is worth mentioning that the Holy Quran avoids direct statements which permit and afford opportunity for non-believers and infidels to deny it in the light of their old non-scientific knowledge (before the age of science). Allah explains that how He raised the heavens and subjected the sun and moon running in its course. Holy Quran mentions in Surah Al-Raad, "Allah is He Who raised the heavens without pillars that you may see; then mastered Himself on the throne of Authority and subjected the sun and the moon, each one running in its course for a term appointed.... (Quran, 13:2)."[38] In fact, the Holy Quran draws our attention to the various signs (of the Creator) embedded in the universe, such as: the vast dimensions of space; the relativity of time; and the alteration of day and night on earth.

The Quranic verses not only speak highly of science and scientists but also draw our attention and direct us to pursue knowledge and discover the secrets of the universe. Allah mentions how He created the universe and expanded it indicating the different galaxies. In Surah Al-Dhariyat of Holy Quran, Allah says, "With power and skill did We construct the heavens and indeed We expand them (Quran, 51:47)".[39] The meaning of this verse is obviously that the universe expands. Galaxies are receding away from us (the Milky Way) at speeds which are proportional to their distances from us. It is only lately in the age of science that it has been discovered that the universe is expanding. Its galaxies are receding at velocities that are increasing with time. This result has been supported by actual observations as well as theoretically according to the Theory of Relativity.

The Holy Quran in Surah Yasin says, "It is not permitted for the sun to reach the moon, nor can the night outstrip the day, and all float along in an orbit (Quran, 36:40)".[40] The Quran again says about the rotation of the sun and moon in an orbit in Surah Luqman, "And He has made subservient the sun and the moon, each one moves swiftly to an appointed time (Quran, 31: 29)".[41] By more clearly defining the myth of the day and night, the Holy Quran in Surah Az-Zumar also mentions, "He maketh night to spherically succeed day and He maketh day to spherically succeed night (Quran, 39:5)".[42] The Holy Quran has its own way of quoting miraculously scientific facts.

The Holy Quran expresses this fact by saying in Surah Al-Israa, "And We have made the night and the day two signs; the sign of the night have We cancelled while We have made the sign of the day lightened, that you may seek bounty from your Master and that you may know the number and count of the years; likewise, everything have We fully explained (Quran, 17:12)".[43] The above verse points to the actual shape of the lowest layer of the atmosphere in which day and night succeed one another, and regards them spherical (shells as we know them now). Evidently, this could be realised only if the Earth is round and rotates in front of the Sun.

The Holy Quran again says in Surah Yunus, "And that is He Who made the sun shine and the moon illuminated and measured out phases for it that you might be aware of the number of years and the count of time; no wise did Allah create this but in Truth and Righteousness, thus He explains His signs in detail for a people who are granted knowledge (Quran, 10:5)".[44] Evidently, this verse refers to the fact that the lunar months and years can be readily counted by observing the phases of the moon.

(ii) *Sky*: About the sky or the atmosphere, the Holy Quran says in Surah Al-Ambiyaa, "And We made the sky (the atmosphere) preserved roof yet they care not to recognise its (multiple) services (Quran, 21:32)".[45] This *Surah* mentions that the earth's roof is its atmosphere. It is composed of a mixture of gases which are preserved and not allowed to escape into space due to the relatively large gravity of the earth. The earth's atmosphere (or roof) renders vital services to living creatures on earth, notably human beings, such as: it contains oxygen necessary for the life of both the animal and plant kingdoms. Secondly, it carries water vapour from the earth's surface up to the upper layers of the atmosphere where clouds form and give rain. This process involves the source of fresh water on earth. Thirdly, it contains carbon dioxide gas, which is absorbed by plants receiving sunlight by day to form sugar, starch, oil etc. Fourthly, sound waves propagate through the air, thus we can hear the voices of each other. Fifthly, daylight is due to the scattering and diffuse reflection of sun rays in the thin air layer extending from the earth's surface up to about 200 kilometres and facing the sun. This layer may be referred to as the (skin-layer).

The lowest sky is the earth's atmosphere, where meteors burn off the excessive heat produced by friction with the air molecules. In this manner, meteors appear like lamps decorating the lowest sky. In this regard, the Quran says in Surah Al-Mulk, "We did decorate the lowest sky with lamps (Quran, 67:5)".[46] About the fall of the meteors and asteroids the Holy Quran again hints in Surah Al-Jinn, "And we (the jinn) pried into the secrets of the heaven, but we found it filled with stern guards (meteorites) and meteors.... And indeed we (the jinn) used to sit there (in the lowest heaven) in selected stations in order to listen; but anyone who listens now will find a meteorite facing him in ambush (Quran, 72:8-9)".[47] It is very amazing that meteorites rarely reach the earth's surface. Most of them, on entering the earth's atmosphere, split into fine dust within the upper layers of the earth's atmosphere. Allah says in Surah Al-Haj, "And He withholds the (lowest) heaven test; it falls on the earth except with His permission. Indeed Allah, regarding people, is most kind and most merciful (Quran, 22:65)".[48]

Allah also explains the oxygen level in the sky that anyone who happens to ascend vertically in the atmosphere will soon feel that he needs more and more air to inhale. This means that he will feel the need to a wider breast to inhale the proper amount of oxygen. This fact was first stated by the Holy Quran, which says in Surah Al-Anam, "That whom He willeth to leave straying He makes his breast close and constricted as if he has to rise up in the sky (Quran 6:125)".[49]

(iii) *Planet and Solar System*: Regarding the planets of our solar system, the Holy Quran says in Surah Al-Saffat, "We have indeed adorned the lowest heaven with an ornament, the planets (Quran, 37:6)".[50] About the number of the planets in our solar system, the Holy Quran indicates in

Surah Yusuf, "Once Joseph said to his father, I have seen eleven planets and the sun and the moon I have seen them prostrating for me (Quran, 12:4)".[51] It is well known that at the time of revelation of Holy Quran only six planets were known including Earth, Mercury, Venus, Mars, Jupiter, and Saturn. Now nine planets are known adding Uranus, Neptune, and Pluto. Recently, the tenth planet, Eris, has also been universally accepted. The eleventh one, Ceres, is also likely to be accepted in the future. So, when at the time of revelation, only six planets, including Earth, were known but Holy Quran indicated eleven planets in our solar system at that time.[52]

It must be kept in mind that the Quranic astronomical or cosmological verses appear in the Quran within the context of the Quranic message. They have a specific vocabulary and are often cited as proofs of the Oneness of God, His attributes, and His inexhaustible knowledge. Nevertheless, they do refer to the physical cosmos and its creation. This physical cosmology is part of the broader creation theme of the Quran, which has multiple meanings that cannot be reduced to the mere physical level.

3. *Earth and Ocean Sciences*

The main basic themes of so-called useful scientific knowledge are miraculously inherent in the text of the Holy Quran for the purpose of human enlightenment. The Quran invites its readers to observe the order and regularities of the universe for the express purpose of understanding the realities that lie beyond the physical realm. This invitation to observe the physical cosmos is often accompanied by an emphatic reminder that observable order and regularities are a sign of the presence of the one and only Creator. The order observable in the physical cosmos is a testimony to Divine omnipotence, power, and wisdom.

In regard to the functions of the universe and chemical actions on the earth, Allah says in Surah Sabaa of the Holy Quran, "He knows all that goes into the earth and all that comes out thereof, all that comes down from the sky and all that ascends thereto, … Who is aware of the unseen, and from Whom is not hidden the least mass of an atom, neither in the heavens nor on the earth, nor is there anything less than that or greater than that but in a conspicuous record…. (Quran, 34:2-4)."[53] Allah further says in Surah Yunus, "... and from your Master is not hidden the least mass of an atom, neither in the earth nor in the heavens, nor is there anything less than that or greater, but in a conspicuous record (Quran, 10:61)".[54]

People generally see that the earth's surface appears flat, in spite of the fact that it is almost spherical. When referring to such apparently described phenomena, the Holy Quran adopts a unique indirect way of expressing things such that it does not urge people to deny the revelation. The Holy Quran mentions the special shape of the earth in Surah Al-Naziaat, "And the earth after that He made as an egg in shape (Quran, 79:30)."[55] The spherical shape of the earth necessitates that when one half of the earth's surface faces the sun, it becomes illuminated. This represents day time. The other half does not face the sun and faces the darkness of space (night). One example is that the earth is round and rotates. The Holy Quran mentions these two facts indirectly in several verses. Long before the age of science, most people regarded the earth's surface as flat as it appeared to them. The fact that the earth is round and that it rotates around its axis and around the sun is also postulated in the Holy Quran, but in an indirect manner.

As a torch bearer for the wise men or scientists on the earth, the Holy Quran always indicates the scope of observation and research. About the fact that the great rivers of agreeable and sweet water when falling into the salty ocean remain the same for a long distance and never mix immediately because of its mixing nature. Holy Quran in Surah Al-Furquan says, "And He (Allah) is the One Who has let free the two seas; one is agreeable and sweet, and the other is salty and bitter. He made a barrier between them, a partition that it is forbidden to pass through (Quran, 25:53)."[56]

It is also a matter of research and observation before the world that the water of two oceans, a recent discovery after many hundred years after the revelation, never mixes with each other and it looks like there is a barrier between them. The Holy Quran in Surah Al-Rahman says, "He (Allah) has let free the two bodies of water meeting together. Between them is a barrier which they do not transgress. Then which of the favours of your Master will you deny? Out of them come pearls and carols. Then which of the favours of your Master will you deny (Quran, 55:19-23)?"[57] Allah mentions in Surah Al-Naml, "He (Allah) Who has made the earth firm to live in and made rivers therein and set thereon mountains and made a separating barrier between the two bodies of water? Can there be another god besides Allah? Nay, but most of them (the infidels) do not know (Quran, 27:61)."[58]

4. *Meteorology (Rain, Wind and Cloud)*

It is a fact that the essentials of natural phenomena such as clouds, rain, wind, daylight etc. have been generally indicated in the Holy Quran before being established by scientists. Regarding the planet Earth and its ability to hold water, distinguishing other planets in the solar system from Earth. Surah Al-Mominun says, "And We sent down water from heaven, according to a precise measure and thereby caused the earth to hold it

(Quran, 23:18)."[59] It means that in the earlier stages of the development of the earth as a planet it could collect water from heavens according to a precise measure, so that it could fulfil its mission and sustain living beings on it. In other words, the quantity of water accumulated on the earth is not haphazardly sent down, in spite of the fact that most of the earth's surface is covered by oceans and seas. By this verse, an observer finds that more than three-quarters of the earth's surface is covered with liquid water. Add to this the huge amounts of ice covering the polar-caps and the extensive amounts of water-vapour found in the atmosphere. No other planet in our solar system could gather comparable amount of water.

However, it is an established fact that water plays the biggest role in the evolution, sustenance, and development of life on earth. Rain is the actual source of fresh water on earth, whether we take this water from rivers, tributaries, wells, or springs etc. This is exactly what the Holy Quran postulates in Surah Al-Waqiaa, "Look at the water you drink. Do you bring it down (in rain) from the nimbus clouds or are We Who let it fall? (Quran, 56:68-69)."[60] The Holy Quran again says in Surah Al-Anam, "And it is He who sends down water (rain) from the sky, with it We produce vegetation of all kinds (plants), from which We produce green (material chlorophyll), out of which We produce heaped grain (Quran, 6:99)."[61]

The Holy Quran explains very widely the process of the formation of winds, clouds, rain, thunder, lightning, snowfall, and ice-cube storming. However, the Holy Quran is known to be the first book that declared and announced the fact that clouds are formed by winds. The Holy Quran in Surah Al-Bakra says, "Assuredly in the creation of the heavens and of the earth and in the alternation of night and day and in the ships which sail on the sea for the benefit of mankind and in the rain which Allah sends down from the sky, thus giving life to the land after its death and dispersing in it all types of beasts and in sending winds in their ways and in the clouds that are made to render service between the sky and the earth are signs for people who understand (Quran, 1:164)."[62]

The Holy Quran again mentions the flora and fauna, mountains, and water on the earth in Surah Fatir, "See you not that Allah sends down water from the sky and then by it We cause the up-growth of fruits of various types and colours; and that of the mountains are tracts white and red of various grades and shades of colour and black intense masses; and of people and beasts and cattle, various likewise are the kinds, but of His servants who fear Allah are only those possessed of knowledge, verily Allah is Mighty, all-Forgiving (Quran, 35:27-28)."[63]

It is well-known now that air contains water vapour, and that when it rises in the atmosphere (by various ways), its temperature decreases adiabatically, and consequently it can no longer hold all its water vapour content. Condensation of the surplus water vapour thus takes place, either in the form of droplets or in the form of ice crystals, according to the prevailing temperature. In this manner clouds develop in the rising air. The Holy Quran in Surah Fatir and Al-Rum says, "Allah is He Who sends forth the winds so they raise clouds; then does He spread them in the sky as He pleases, and He makes them partially thick so that you see the rain coming from inside them; then when He causes it (i.e. the rain) to fall upon whom He pleases of His worshipers, they rejoice (Quran, 35:9&30:48)."[64] It is again mentioned in Al-Nur, "Seest thou not that Allah drives along some clouds then joins them together then makes them into heaps of cumulus, then wilt thou see rain issuing from their inside and He sends down from the sky mountain masses of cumulus wherein is hail, afflicting therewith whom He pleases and turning it away from whom He pleases, the vivid flash of its lightning (whose electric charge is due to hail) almost takes away the sight (Quran, 24:43)."[65]

It is well-known that for rain to fall from clouds, the winds forming the clouds must continue to seed them with water vapour and the so-called nuclii of condensation, which are mainly sea-salts. We know now that rain is not stored in the sky but a water cycle. In this respect, the Quran in Al-Hijr says, "And We send the winds seeding the clouds, thus causing water to descend from the sky, therewith providing you with it to drink; and you are not the owners of its stores (Quran, 15:22)."[66]

5. *Geology and Mineralogy*

Geology or mineralogy, a science that deals with the dynamics and physical history of the earth, the rocks of which it is composed, and the physical, chemical, and biological changes that the earth has undergone or is undergoing, is a branch of science to which attention was directed by the Holy Quran itself. What was observable on the earth was a vast system of change over long periods of time, and this system was taken as a sign of Allah. The Muslim scientists studied various geological processes, such as weathering, erosion, and transportation, the stratigraphic arrangement of strata, and long geological spans of time within the context of the Quranic descriptions of creation. Many Muslim scientists, such as Al-Biruni, Al-Kindi, Ibn Sina, Al-Masudi, and Al-Idrisi, have left valuable works on geology and mineralogy. All of these works reflect a worldview anchored in the Quran and a belief system that takes the earth as a sign of the Creator.

The Holy Quran says of the expansion and drifting of the land slots and mountains and their merging into other land slots, in Surah Al-Naml,

"And you see the mountains and think them firmly fixed (stationary), but they pass away as the clouds pass away (Quran, 27:88)."[67] It is a true fact that the name of the ultimate small size of material bodies is "atom". It is also true that the atom could not be divided into smaller constituents. Its divisibility was realised only in the age of science, but the Holy Quran mentions it many hundred years before.

The Quran also gives more emphasis on the knowledge of geology and mineralogy. In a later stage of the revelation, the Quran, for example, asks the people to travel here and there to see how creation began. Scientifically, notably in natural history, this is, as we are aware now, in order to gather and collect various fossils, and it is accepted that the earth's history is written within the layers of its crust. Special attention was paid to the formation of rocks, mountains, the course of rivers, natural processes, minerals and lapidaries. Various stones and gems were studied for their medicinal properties, and many stones and minerals were converted into digestible form and were thus incorporated into the Islamic pharmacological tradition.

In the Quranic Surah Al-Ankabout, Allah says, "Go and travel through the Earth and see how Allah did originate creation (Quran, 29:20)."[68] We can explain this verse scientifically by saying, "go and travel through the earth in order to study its crust, and to collect fossils from here and there and the remains of organisms that lived throughout the ages." The verse also indicates that there is no evolution from one species to another, but mere development due to acclimatization, such as a change in colour, or adaptation due to the environment. Man was created as a man and remained as a man and study of these things leads to an understanding of the origin of life. In other words, evidence of things throughout the ages is recorded on the rocks of these ages in the form of fossils, or harried things that can be carefully studied and traced back.

6. *Biology (Zoology and Botany)*

About biology or zoology (human anatomy and morphology, human life cycle) and botany (flora and fauna), the Holy Quran has very clearly described many details in its verses. The Holy Quran mentions a number of animals and birds by name and speaks of their benefits to humanity. Chapters 2 and 16 of the Holy Quran are named after the cow and the honeybee, respectively. Horses are specifically mentioned in many verses. For instance the first five verses of the hundredth chapter, for instance, describe charging horses in magnificent rhythm. Eating the meat of certain animals is considered unlawful in Islamic law. Many plants and fruits are also mentioned in the Holy Quran. These verses of the Quran provide a certain theoretical framework for the cultivation of zoology, botany, and other related sciences.[69]

More than 1400 years ago, the Holy Quran miraculously declared that prenatal human development takes place in various stages. Those stages were discovered mainly during the middle of the nineteenth century, after the discovery of the ovum. Previously, scientists had completely ignored the role of the woman in the development of the embryo. After thorough study, scientists finally became aware of the fact that embryonic development occurred in stages, which required accurate descriptions. The Holy Quran says in Surah Al-Muminun, "And we did create man from an evolutionary quintessence of clay. We then made him a fertilised drop attached firmly to a place of settlement. Then We made of the fertilized drop a leech-like structure, and then We made of the leech-like structure a chewed-like piece of flesh, then We turned the chewed-like flesh into bones (skeleton), then We clothed the bones with flesh (muscles), and then We caused him to develop and come into being (human) (Quran, 23:12-14)."[70]

In Surah Al-Hajj, the Holy Quran gives the knowledge of the whole human cycle and the day of resurrection, "O mankind if you are in doubt about the resurrection, (consider) that We have created you out of dust, then out of sperm, then out of a leech-like clot, then out of a morsel of flesh, partly formed and partly unformed, in order that We may manifest Our power to you; and We cause whom We will to rest in the wombs for an appointed term, then do We bring you out as babies, then foster you that you may reach your age of full strength, and some of you are called to die, and some are sent back to the feeblest old age, so that they know nothing after having known (something); and further, you see the ground barren and lifeless, but when We let rains fall on it, it stirs and swells, and puts forth every type of beautiful pairs of growth. This is so because Allah is the Reality, and it is He Who gives life to the dead, and it is He Who has power over all things. And verily the hour (or the last day) will come, there can be no doubt about it, and about the fact that Allah will raise up all who are dead (Quran, 22:5-6)."[71]

7. *Neurology*

Referring to the activities of the brain at the time of the death of the human body, the Holy Quran says, "At the moment of death, everyone knows his or her destiny; heaven or hell. For the disbelievers, death is a horrible event; the angels beat them on the faces and rear ends as they snatch away their souls (Quran, 8:50, 47:27 & 79:1)."[72] The verses may be described as the Holy Quran tells us that everyone is shown their previous activities (*Amaal Naamah*) just before their death. Naturally, his brain reacts to that, which left effects just before and after the departure of his soul. It is undoubtedly a neurological process and not clearly visible to the others, but it left a little effect on the body of the dying person.

A study by Dr. Ajmal Zemmar, a neurosurgeon at the University of Louisville, United States, published in the journal, "*Frontiers in Aging Neuroscience*" on 22 February, 2022, proves the Quranic verse true. The study suggested that life may actually flash in front of our eyes before death. Scientists have for the first time recorded the brain activity just 15 minutes before a human death, and it revealed astounding results. Researchers found "rhythmic brain wave patterns" near the time of death with similarities to those experienced during dreaming and meditation. They recorded there was an increase in "gamma oscillations", which occur during dreaming and memory retrieval.[73]

Dr. Ajmal Zemmar said, "In the 30 seconds before the patient's brain stopped receiving blood, his brainwaves followed the same patterns as when we carry out high-cognitive-demanding tasks, like concentrating, dreaming, or recalling memories." He continued, "This could possibly be a last recall of memories that we have experienced in life, and they replay through our brain in the last seconds before we die," Dr. Zemmar clarified, "Through generating oscillations involved in memory retrieval, the brain may be playing a last recall of important life events just before we die, similar to the ones reported in near-death experiences."[74]

8. *Calendar and Mathematics*

It has been regarded that mathematics, or arithmetic has been the first step of scientific development. The Holy Quran mentions the number of days in the solar and lunar months and also fixes twelve months in a year. In Surah Al-Tawba, Allah says, "For sure, the number of months (per year), as regarded by Allah, is twelve months in the Book of Allah on the day He created the Heavens and the Earth (Quran, 9:26)."[75] Allah also clarified the exact difference in days per month and year between the solar and lunar calendars. The Quran transfers 300 solar years to 309 lunar years. The Surah Al-Kahf says, "They remained in their cave for three hundred years, a number which has to be increased by nine years (Quran, 18:25)."[76] This verse was the source of scientific research which differentiated the solar and lunar months and year.

In its series of mathematical knowledge, the Holy Quran especially mentions the decimal system, or 1/10, in its verses while commanding the people. In Surah Sabaa Allah says, "And those who were their predecessors rejected the Truth even though they did not receive one tenth of what We had granted (Quran, 34:45)."[77] The Holy Quran again mentions in Surah Al Nisaa, "Allah directs you as regards your children's (inheritance), for the male a portion equal to that of two females; but if only women more than two their share is two thirds of the inheritance, and if only one, her share is one half; for parents, a sixth share of the inheritance is for each if the deceased left children; if he has no children and the parents are the (only) heirs, the mother takes the third, but if the deceased has brothers (or sisters), the mother takes the sixth. The distribution in all cases is (always) after the payment of legacies or debts (Quran, 4:11)."[78] It is very interesting that about two hundred years after the revelation of the Holy Quran, the world focused on science and finally discovered the decimal system and its implementation.

9. *Day of Resurrection*

Another unsolved challenge before the scientists is the cosmic revolution, which will destroy everything, as mentioned in the Holy Quran, *Qayamah* or Day of Resurrection. Quite amazingly, the Holy Quran is the only source of this knowledge and a great matter of research for scientists. It is said that all the natural processes will cease to function. This cosmic revolution of the *Qayamah* will be so great and beyond imagination. The physical world will vanish, revealing the true and real spiritual truth. Allah says in Surah Fatir, "It is Allah Who sustains the Heavens and the Earth lest they annihilate, and if they should annihilate there is none, not one, who can sustain them thereafter (Quran, 35:41)."[79] Allah again says in another verse of Surah Ibrahim, "One day the earth will be changed to a different earth, and so will be the heavens, and men will be marshalled forth before Allah, the One, the Irresistible (Quran, 14:48)."[80]

On reviewing about a quarter thousand verses of the Holy Quran, the Muslim intelligentsia and even non-Muslim stalwarts come to the conclusion that the Quran is really a miraculous book of knowledge and a torch bearer to mankind from the day of its revelation to the day of resurrection. This knowledge contains not only religious matters but almost all branches of science and repeatedly enforces mankind to search and research the creation of Almighty Allah to gain further knowledge. By following the Quranic guidance, many scientists of the Islamic Empire glorified and adorned the world with scientific discoveries and experiments. For the first time, the practical, laboratory and observatory were utilised for scientific research, which were undoubtedly essential parts of scientific research and development.

NOTES

1. A Prophet was a chosen human being who had been told by God something to convey to the people. Each country and each age has had its own prophet. Each prophet preached to his own people in his own language. The language, the people, and the period may have been different, but the religion was the same in every case. In this process, Allah "taught man that which he knew not" by means of a chosen human being. The Prophet Mohammad (pbuh) was the messenger of Allah who had been told the message of Islam and the words of its most important embodiment, the Quran, and who then faithfully conveyed this message to his people.
2. The term "pbuh," meaning "Peace Be Upon Him" to indicate the respect paid by Muslims to the Prophet Mohammad (pbuh).
3. *Al-Quran.*
4. Dahlan, Winai (2015). *Muhammad: The World's Great Scientist*, Bangkok, p. 80.
5. *Al-Quran.*
6. *Ibid.*
7. Nasr, Seyyed Hossein (2010). *In Search of the Sacred: A Conversation with Seyyed Hossein Nasr on his Life and Thought*, Santa Barbara, California, p. xiii.
8. *Al-Quran.*
9. *Ibid.*
10. *Ibid.*
11. *Ibid.*
12. *Ibid.*
13. *Ibid.*
14. *Ibid.*
15. *Ibid.*
16. *Ibid.*
17. *Ibid.*
18. *Ibid.*
19. Ar-Rafai, Shaikh Mohammad Nasib (1998). *Tafsir Ibn Kathir*, (Part-I), London, p. 214.
20. *Al-Quran.*
21. *Ibid.*
22. *Ibid.*
23. *Ibid.*
24. *Ibid.*
25. *Ibid.*
26. *Ibid.*
27. *Ibid.*
28. *Ibid.*
29. *Ibid.*
30. *Ibid.*

31. *Ibid.*
32. *Ibid.*
33. *Ibid.*
34. *Ibid.*
35. *Ibid.*
36. *Ibid.*
37. *Ibid.*
38. *Ibid.*
39. *Ibid.*
40. *Ibid.*
41. *Ibid.*
42. *Ibid.*
43. *Ibid.*
44. *Ibid.*
45. *Ibid.*
46. *Ibid.*
47. *Ibid.*
48. *Ibid.*
49. *Ibid.*
50. *Ibid.*
51. *Ibid.*
52. El-Fandy, M.G. (1992). *Islam and Science*, Vol. 6, Cairo, p. 24.
53. *Al-Quran.*
54. *Ibid.*
55. *Ibid.*
56. *Ibid.*
57. *Ibid.*
58. *Ibid.*
59. *Ibid.*
60. *Ibid.*
61. *Ibid.*
62. *Ibid.*
63. *Ibid.*
64. *Ibid.*
65. *Ibid.*
66. *Ibid.*
67. *Ibid.*
68. *Ibid.*
69. Iqbal, Muzaffar, *op. cit.*, p. 67.
70. *Al-Quran.*
71. *Ibid.*
72. *Ibid.*
73. *Frontier Aging Neuroscience*, 22 February, 2022, https://doi.org/10.3389/fnagi.2022.813531

74. *https://www.timesnownews.com/viral/life-might-actually-flash-before-our-eyes-as-we-die-study-article-89822967 Al-Quran.*
75. *Al-Quran.*
76. *Ibid.*
77. *Ibid.*
78. *Ibid.*
79. *Ibid.*
80. *Ibid.*

3
Contribution of *Hadith* in Development of Science

Hadith is an Arabic term which literally means a narrative, report or account. *Hadith*, or Sunnah, encompasses all the traditions and practices of the Prophet Mohammad (pbuh), including his deeds, sayings, ideas, permissions, and disapprovals. In other words, it may also be said that a *Hadith* is an anecdote reporting something that the Prophet said, did, or did not do, that someone said or did in his presence without his objecting, or that one of his knowledgeable Companions said or did. These were recorded simultaneously as the Quran. Along with the Quran, the Sunnah makes up the two primary sources of Islamic theology and law. The *Hadith*, however, were in the thousands, and only a portion of the Ahadith of the Prophet existed in written form during his lifetime.

Prophet Mohammad's (pbuh) followers largely gathered the *Hadith* while only a small part was done by the state. As for the collection of *Hadith*, or the Prophet's traditions, including sayings, deeds, recognition, refusal, personality as well as routine from his birth to death, it continued for centuries after the Prophet had died. The compilation of *Hadith* led to the emergence of a rich tradition of *Hadith* studies, which in turn called for the development of exacting methodologies and techniques to authenticate, index, and cross-reference a very large number of individual sayings of the Prophet. The record might be in a formal written or memorised report and classified as "authentic" (*Sahih*), "good" (*Hasan*) and weak (*Zaeef*), which was argued and categorised based on reliability.

These Ahadith (plural of *Hadith*) contributed to the academic development of both Arab and non-Arab Muslims to uplift their society and create fields of science, from which the prosperous Islamic civilization flourished. Thus, seeking *Hadith* was an interesting part of Islamic history. This intellectual awareness was the thing that gradually civilised Arab society. Such dedication stirred up the Muslim community's enthusiasm to study, research, and build a society of wisdom. To conclude, the collection of the Quran and *Hadith* played a major part in building Islamic civilization. Muslim scholars collected hundreds of thousands of Ahadith during the first three centuries of Islamic history, compiling them according to various principles. During the course of the ninth century, *Hadith* gained acceptance as a source of Islamic law, largely supplanting more diffuse conceptions of Sunnah.[1]

During the time of the Caliphate, the process of collecting the *Hadith* or Traditions of the Prophet was started, and the official recording of the Prophet's *Hadith* was conducted at the time of the Umayyad Caliph Umar Ibn Abd al-Aziz. Furthermore, *Hadith* collection continued under the Abbasid Caliphs. The study and research of Prophetic Traditions, in addition to the Quranic verses, opened the door of knowledge and learning and, ultimately, the study of science and philosophy. The scholars found that the Prophet Mohammad (pbuh) had elaborated the teachings of the Quran and proclaimed that seeking knowledge should be the faith of every Muslim man and woman, and he or she must continue to acquire knowledge from the cradle to the grave and follow it wherever they can acquire it.

We could see that in *Hadith* there is the chain of authorities, e.g. Ibn 'Umar, who was a young man during the Prophet's later years, heard him say, "Islam is built on five things..." and then told the story to Tkrima, who told it to Hanzala, who told it to 'Ubayd Allah, who told it to Bukhari, each of the five men being careful to pass on the exact wording of the story as he heard it from the chain of previous authorities. Imam Nasai (829-915), Ibn Maja (824-887), Imam Tirmidhi (824-892), Imam Muslim (821-875), Abu Dau'd (817-889), Al-Bukhari (810-870), and hundreds of other scholars collected *Hadith* for knowledge, learning, and guidance of the Muslim world.

The sciences of the Quran were the first to appear in the Islamic intellectual tradition. This was followed by the sciences dealing with the life and sayings of the Prophet of Islam. Thus, when the scientific and philosophical traditions emerged in Islamic civilization, they emerged from within a specific intellectual context shaped by the Quran and *Hadith*. Following the Holy Quran and *Hadith*, the early Muslims spread throughout the world in quest of knowledge and learning. The Prophet Mohammad (pbuh) said: Allah, His angels and all those in the heavens and on earth, even ants in their hills and fish in the water, call down blessings on those who instruct others in beneficial knowledge (*Al-Tirmidhi*, 422).[2]

In the preceding Ahadith, we can see that Prophet Mohammad (pbuh) placed a high value on education, knowledge, and scientific research. He was a peace lover and always tried to avoid armed conflict until forced to fight. The Prophet and his fledgling group of followers in Medina were forced early on to fight for their survival against attacks from the elites of Makkah. A series of skirmishes culminated in the battle of Badr outside Makkah, the first battle for Islám, in which a small Muslim army triumphed over a much larger Makkan army. For the Muslims, this victory was powerful proof that God was on their side, and a large number of people accepted Islam.

The Quran and *Hadith* together are the fountainhead of all that is Islamic in whatever domain it might be. More specifically, these sources have played a double role in the creation and cultivation of the sciences. First of all, the principles, not, of course, the details, of all sciences are considered by Muslims to be contained in the Quran, and there is an esoteric interpretation of the Holy Book which makes possible the unveiling of its mysteries and penetration into its inner meaning wherein the principles of all the sciences reside. Secondly, the Quran and the *Hadith* have created an atmosphere for the cultivation of the sciences by emphasising the virtue of pursuing all knowledge that is, in one way or another, a confirmation of Divine Unity. Therefore, a whole metaphysics and cosmology have issued from the bosom of the Quran and the *Hadith* and have acted as the basis upon which all the Islamic sciences have been constructed.[3]

The knowledge of Quran and *Hadith* have also created a particular atmosphere which has fostered and encouraged all intellectual activity that is in conformity with the spirit of Islam as reflected in the Quran and the *Hadith*. Consequently, Muslim scholars received this fact clearly and then contributed uniquely to human history in terms of science in countless volumes that can never be presented in a single study. The acquisition of knowledge is virtuous; it ennobles humanity and it serves its needs. It is this religious obligation that provides a nexus between Islam and the quest for scientific knowledge. The Muslim intellectuals emerged as undisputed leader of scientific knowledge and discoveries continuous for seven centuries and their work still benefit the modern civilized world. These crowning successes were largely due to tremendous ideological motivation provided by Islam for the study of natural phenomenon and pursuit of empirical knowledge.

Scientific Knowledge and Hadith

The Arab people in the Arabian Peninsula before the birth of Prophet Mohammad (pbuh) were uncivilised and underdeveloped. But after Mohammad (pbuh) came and spread Islam, only by accepting Islam did the uncivilised community turn their society into a leading group of people who could govern and bring prosperity to the old civilization. The development began after the appearance of the Prophet and his introduction of Islam, with the Quran as the main mechanism. The Prophet spent 23 years spreading Islam, including 13 years in Makkah and 10 years in Madinah after the migration.[4]

Prophet Mohammad (pbuh) was not a scientist as perceived in modern times, but he was the type that built foundations for far more advanced science in the future. Through his teachings, Arabs were exposed for the first time to a set of new ideas on creation, the Supreme Creator, the purpose of life on Earth and in the hereafter, the need for a code of ethics in private and public life, the obligation to worship the one and only Lord Almighty of the Universe (Allah) through ritual prayers on a regular basis and sessions of remembrance (*Dhikr*, plural *Adhkar*) or meditation, and to pay homage to religious and political leaders. All this was new to the Arabs. The whole package of Islamic teachings was propagated by the Prophet and accepted by his fellow Arabs.

The Prophet urged people to get curious and learn new ways of thinking, which helped them bring their thoughts into daily practice. All the later leaders of the Muslim world followed in his footsteps, providing support for scholars, which contributed to the development of basic sciences including mathematics, chemistry, biology and medicine. Consequently, the Muslim world turned into a source of learning as well as research. The development that stemmed from the Prophet was notably crucial to building the society of wisdom, especially in science and technology.

Seeking Scientific Knowledge

Knowledge or seeking scientific knowledge is an essential part of Islam. One cannot think of a better description of knowledge, its needs and its attributes than what the Prophet Mohammad (pbuh) has explained. Like the other prophets, Prophet Mohammad (pbuh) was sent to spread the message of Allah among the people for the benefit of mankind. He used Islam to nurture the people, teaching them about hygiene, sanitation, and balance in society, and not to take advantage of others. Muslims learn rules from religious teachings such as social etiquette, leadership, and obedience to leaders through prayers, sacrifice, sympathy, tolerance through observing fasting, sharing, a fair society through *Zakah* (charity), harmony, equality, and how to build a multiracial community through building the nation and the Hajj pilgrimage. Apparently, teaching in Islam is done through practical demonstration.

A new Muslim society that emerged since the beginning of Islam strongly supported the spread of education. An important incident that serves as a turning point for the Arab people was that when Prophet Mohammad (pbuh) was 40 on 21st Ramadan, the ninth month according to the Arabian calendar that was on the 21st August in the year 610, he received *Wahi* (revelation), the first revelation, in the Cave of Hira on the mountain of Jabal Al-Noor, three kilometres north-east of Makkah. In the month of Ramadan, occasionally the Prophet would retreat to the cave with clothes and some dried fruits for several nights of seclusion and prayer. On one such occasion, while he was in contemplation, the angel appeared to give him the first revelations from Allah.

For the first revelation that dawned on the Prophet started with the word *Iqra*, which means "read". Allah repeats the word "read" three times, then says that He the Almighty has taught mankind; taught what mankind knew not. Allah then focuses on the importance of the pen in the acquisition and spread of knowledge. Prophet Mohammad (pbuh) always encouraged Muslims to acquire knowledge as clear by his famous statements: "Seek knowledge even if it is in China"; he said "Seek knowledge from the cradle to the grave"; and "If anyone travels on a road in search of knowledge, God will cause him to travel on one of the roads of Paradise". Based on Prophet Mohammad's (pbuh) prophetic image, the Western people viewed his great efforts to create a glorious Muslim society, saying, "When a mountain does not come to Mohammad, he has to approach it".[5]

There are no definite sources claiming that science was prevalent during the time of the Prophet. Despite the paucity of early sources, we can confidently trace two branches of science—medicine and astronomy, to the days of the Prophet himself. Prophet himself gave much emphasis to medicine and psychology. The words and practices of the Prophet formed the basis of practical psychology, which was greatly advanced by Muslim scientists and scholars. Prophet Mohammad (pbuh) recommended specific supplications for the sick. It was the practices of Prophet Mohammad (pbuh) to visit the old and sick and give them hope and joy. Practical psychology was also shaped by the Quranic verses dealing with the relationship between the human condition and the Creator. In the Holy Quran it is mentioned that, "Indeed, the hearts receive tranquility by the remembrance of Allah" (Quran, 13:28).[6]

The Prophet's work was performed with rationale and reason, aiming for concrete results. Many of his explanations of the Quranic verses were related to scientific knowledge, including biology, geology, astronomy, and mathematics. Modern scientists admit that both the Quran and *Hadith* or Sunnah are surprisingly accurate after examinations and experiments. The knowledge he passed on and the legacy he left for the Muslim world were built upon and inspired many globally recognised Muslim astronomers, chemists, physicists, mathematicians, engineers and medical doctors.

Within a decade of death of the last Prophet Mohammad (pbuh), the Islamic state of Madinah politically gained momentum and converted to an Empire. The army of the Muslims conquered all of the Arabian Peninsula. Within a century, the Islamic Empire had spread from Al-Hamrah in Spain to the borders of China. In Islam, science, theology, and philosophy are unified. Muslims were commanded by Allah Almighty to study, seek knowledge, learn, and benefit from others. It was the Holy Quran and the Sunnah that inspired the Muslims to great heights in the sciences, medicine, mathematics, astronomy, chemistry, philosophy, art, and architecture.

Up to the time of the Islamic dispensation, science and literature possessed no votary in the Arabian Peninsula. However, in the very first verse of the Holy Quran revealed to the Prophet of Islam, the importance of reading, writing, and acquiring knowledge was expounded in a very clear, decisive, and direct manner. The words of the Prophet gave further impulse to the awakened energies of the race. Even within his lifetime, he formed the nucleus of an educational institution, which in later years grew into universities at Bagdad and Salerno, as well as at Cairo and Cordova.

The Prophet's devotion to knowledge and science has distinguished him from all other teachers, and brings him into the closest affinity with the modern world of thought. The city of Madinah, the seat of the theocratic commonwealth of Islam, had, after the fall of Makkah, become the centre of attraction, not only to the hosts of Arabia, but also to inquirers from abroad. Persians, Greeks, Syrians, Iraqis, and Africans of various hues and nationalities from the north and west flocked here. Some, no doubt, came from curiosity, but most came to seek knowledge and to listen to the words of the Prophet of Islam.[7]

While preaching the value of knowledge, Prophet Mohammad (pbuh) said, "Acquire knowledge, because he who acquires it in the way of Allah performs an act of piety; who speaks of it, praises Allah; who seeks it, adores Allah; who dispenses instruction in it, bestows alms; and who imparts it to its fitting objects, performs an act of devotion to Allah. Knowledge enables its possessor to distinguish between what is forbidden and what is not; it illuminates the path to Heaven; it is our friend in the desert, our society in solitude, our companion when we are alone; it guides us to happiness; it sustains us in misery; it is our ornament in the company of friends; and it serves as an armour against our enemies. With knowledge, the servant of Allah rises to the heights of goodness and to a noble position, is associated with sovereigns in this world, and attains to the perfection of happiness in the next."[8]

To boost the Arabian Muslims' ability to gain knowledge, Prophet Mohammad (pbuh) mostly gave reference to the Persians, who were masters in acquiring knowledge and education in those days. During the heydays of Islamic civilization, the Persians were the first who became masters of knowledge, literature, and science, and the saying of the Prophet was verified: "If knowledge were attached to the end of the sky, some amongst the Persians would have reached it".[9]

Prophet Mohammad (pbuh) gave much importance to science, knowledge, and learning and frequently said, "One hour's meditation on the work of the Creator (in a devout spirit) is better than seventy years of prayer." "To listen to the instructions of science and learning for one hour is more meritorious than attending the funerals of a thousand martyrs, more meritorious than standing up in prayer for a thousand nights;" "To the student who goes forth in quest of knowledge, Allah will allot a high place in the mansions of bliss; every step he takes is blessed, and every lesson he receives has its reward;" "The seeker of knowledge will be greeted in Heaven with a welcome from the angels;" "To listen to the words of the learned, and to instil into the heart the lessons of science, is better than religious exercises, ... better than emancipating a hundred slaves;" "Him who favours learning and the learned, Allah will favour in the next world;" "He who honours the learned honours me;" "Eminence in science is the highest of honours;" "He did not die, who gives life to learning;" "The greatest ornament of a man is erudition."[10]

Prophet Mohammad (pbuh) said about seeking and spreading knowledge, "Allah, His angels and all those in the heavens and on earth, even ants in their hills and fish in the water, call down blessings on those who instruct others in beneficial knowledge".[11] Naturally, such sentiments on the part of the Master and the Chief of the disciples gave rise to a liberal policy and animated all classes with a desire for learning. Consequently, Muslim scholars understood this fact clearly and then contributed uniquely to human history in terms of science in countless volumes that can never be presented in a single study.

Earth and Cosmos

Besides seeking knowledge of every subject, there are many traditions in revealing the scientific fact of the miracles of the Prophet's *Hadith*. The shape and expansion of the earth is one of the most important *Hadiths* that contains a scientific explanation. Related to this subject, Prophet Mohammad (pbuh) said that Islam would spread as far as night and day, covering every area on the face of the earth.[12] In this piece of *Hadith*, there are two subject matters explained, one about the Muslim Ummah (community), and the other about time. People in the past could not understand this *Hadith*, but modern science has proven its precision. In the case of the Muslim community population statistics have confirmed that Muslims are present in 196 countries in the world. Every single country has a number of Muslims. It is also estimated that in 2025, Muslims will make up one-third of the world population. In other words, Muslim will comprise the majority of the world's population.

In terms of the overlapping of time, the *Hadith* indicates that day and night span the earth, which implies that day and night happen simultaneously on the globe. However, people in the Prophet's time did not understand the true meaning of this, because they had no idea how day and night could occur at the same time, thinking that the globe was flat. Regarding this issue, the Prophet rephrased a Quranic verse (39:5), which describes the world having day and night concurrently, "He created the heavens and the earth in truth. He wraps the night over the day and wraps the day over the night and has subjected the sun and the moon, each running fits course for a specified term. Unquestionably, He is the Exalted in Might, the Perpetual Forgiver." In addition, in Quran (79:30) Allah says, "And after that, He spread the earth."

In addition, Prophet Mohammad (pbuh) said later that the imposition of prayer and fasting times should be in accordance with local times, which was likely to emphasise that different areas on the earth had different days and times. Nonetheless, the people in his time did not understand this concept, and nobody could have imagined the world having time differences.[13] It was stunning how an illiterate man knew the facts that appeared in the *Hadith* regarding the earth's round shape as well as the time differences. The *Hadith* intensified the correctness of the information given in the Quran, the undisputed miracle of the Prophet.

By taking the *Hadith* narrated by Ibn Abbas, that the Prophet was asked about the sinking of celestial bodies (celestial), and the rising of celestial objects. Then the Prophet replied, "he remained in his place, did not move and shift, he drowned for one people and published for one other people, and at the same time, one people declared sinking while others stated rising."[14]

Following the footsteps of Prophet Mohammad's (pbuh), the Muslim scientists discovered that the earth is round and they did 50 centuries before the European discoveries. In 885, a Muslim scientist named Ibn Khordadbeh (820-912) reported that the earth was round like the egg yolk inside the whole egg, which represented the universe. In 903, Ahmad Ibn Roste Esfahani or Ibn Rustah gave his opinion that Allah created this universe in the shape of a ball with nothing inside other than the earth and other round stars in its orbit. Moreover, there were many other Muslim scientists who confirmed that the world is round, including Ahmad Ibn Mohammad Ibn Abdu Rabbi (860-940) and the astronomist Abi Ubaidah bin Ahmad. The earth is a circular planet, according to all of them.[15]

Of course, some celestial bodies disappear from the earth's inhabitants in one region and rise in another region. Objects have been circulating on certain lines or orbits, so they do not move or move in the slightest. It is not possible, except if the earth is round or elliptical, because a stretch without an edge is a rounded thing, and the only endless form is a round shape. It also confirms that the earth is round, revolves around the sun, and rotates on its axis.

Soil

Many of the sayings of the Prophet which have been scientifically proven are about the soil. He said in a report of *Hadith* narrated that Jabir bin Abdullah said: "The Messenger of Allah said, 'The earth (soil) has been made for me a place of prostration and a means of purification..."[16] In regard to the latter, Islam prescribes that soil be used as a cleaning material for things that are particularly ritually unclean, called *Najis Mughallazah*, such as dog saliva. Many Islamic scholars classify pigs as the worst Najis, and this type of impurity can only be cleansed with soil. Prophet Mohammad (pbuh) further said: "When a dog licks anyone's vessel, wash it seven times, the first time with soil and water."[17]

Professor Jo Handelsman, a microbiologist at Yale University, said that she never regarded soil as an unclean material. She acknowledged that many bacteria in soil contained antimicrobial genes. However, her research team never found soil bacteria that resists antibiotics used in humans. In fact, pathogenic bacteria are not from soil. In addition, Dr. Handelsman and her colleagues discovered that soil was a source of microorganisms to create new antibiotic drugs that deal with antimicrobial resistance. In addition to not being troublesome, soil is also a drug that tackles such problems, she concluded.[18]

Many scientists have found that some soils have the ability to kill almost all types of pathogenic bacteria. According to studies of the antibacterial activity of soil, it was found that putting pathogens in a soil solution for 24 hours would destroy all of them, whereas dipping them in other solutions would allow them to split 45 times.[19] This information led to the conclusion that most antibiotics found in soils were created by microorganisms dwelling in the soils. These facts are undeniable proof that Prophet Mohammad's mention of the use of soil for purification was factual. He said it long ago and the proof was discovered 1400 years later by modern scientists.

Medical Science

The verses of the Quran and the practices of the Prophet were used by Muslim scholars to develop a comprehensive framework for the practice of sciences related to health, preventive medicine, and psychological well-being. Studies in the branch of medicine were accompanied by an effort to preserve the exact words of the Prophet. His sayings dealing with health, sickness, hygiene, and other issues related to medicine contain specific references to diseases such as leprosy, pleurisy, and ophthalmic diseases. He recommended remedies such as cupping, cautery, and the use of honey and other natural substances. The Ahadith on medical issues were systemized by religious scholars who were often practising physicians.

Thus, literature on *Tibb al-Nabawi* (Prophetic Medicine) is a distinct genre in Islamic medical sciences and there are numerous works dealing with various aspects of this tradition. One of the most celebrated works of this type is the *Tibb al-Nabawi* of Ibn Qayyim al-Jawziyya (1292-1350). In his book, Ibn Qayyim al-Jawziyya provides general principles of health and sickness, reflects on the relationship between medicine and religion, enumerates the sayings of the Prophet concerning medicine, and discusses the role of Divine Revelation in medicine. He also specifically mentions various remedies recommended by the Prophet.[20]

Another related aspect of this tradition of the Prophetic medicine is the body of literature dealing with pharmacological studies on various herbs and other natural substances, used or recommended by the Prophet. A whole branch of scientific research has been inspired by the teachings of the Prophet on the usage of these substances and these studies continue to this day.[21] Sayings of the Prophet dealing with health, sickness, hygiene, and specific diseases and their cures were compiled and systematized by later generations of Muslims, and this body of literature provided the foundation for a specific branch of medical science in Islam.

Another *Hadith*, regarding the heart as the centre of the body, is the Prophet's saying, "Indeed in the body, there is a clot of blood which if it is healthy and good, then the whole body is good. Conversely, if he is sick, then his whole body hurts. Know that a blood clot is the heart."[22] It is said that if the heart is attacked by a disease, then the whole body will be damaged. That is because the heart's function is to pump dirty blood from the right ventricle into the lungs to carry out blood oxidation. Then the oxidised blood is returned to the heart in the left ventricle to be spread throughout the body. That is the process of working out the material of the heart. Its job is to supply the cells forming the human body with oxygen gas and food juices.

Prophet Mohammad (pbuh) also encouraged people to use medicine for their illnesses. His message encouraged Muslim scientists to search for medicines for each and every illness. In Islam, the human body is a home of indebtedness; in what way it functions, by what method to keep it fresh and safe, in what manner to prevent diseases from attacking it, or what remedies to seek for those diseases, have been important issues for Muslims. Prophet Mohammad (pbuh) himself said, "Allah created no illness, but established for it a cure, except for old age. When the antidote is applied, the patient will recover with the permission of Allah."[23] So the interest in medicine goes back to the time of the Prophet Mohammad (pbuh), who once said that "there is always a cure that exists for every disease." These were solid inspirations to boost Muslim scientists' ability to discover, progress, and spread empirical laws.

For the medical precautions and preventive measures of diseases, the Prophet Mohammmad (pbuh) has given more emphasis on cleanliness, in terms of both personal hygiene and the living environment. Thus the Prophet (pbuh) declared that "cleanliness is a part of the faith (of every Muslim)."[24] He also said that "Allah does not accept *Salah*, a ritual prayer, that is not preceded by ablution."[25] Personal cleanliness, clean clothes, and ablution are parts of the daily observances of all practising Muslims, and there is much attention to details in the Quran and Sunnah on such matters, as the requirements of ablution and bathing, regular brushing of one's teeth, washing of hands before and after meals, and observance of hygiene when drawing and drinking water from wells and springs.[26]

Technology

Prophet Mohammad (pbuh) was bitterly against the scientific invention of destructive technology. For him, seeking knowledge and scientific research should be based on the welfare of humanity and all living and non-living creatures. Even the Prophet forbade acts of destruction and cruelty in normal day-to-day life. He forbade the setting of fire to an anthill by one who might have been stung by a single ant. He also forbade the killing of bees and captured livestock, as killing them is a form of mischief. He once ordered a man who had taken away the nestlings of a bird from their nest to return them to their nest. Muslim jurists have consequently held that destroying a living creature that does not pose a threat to one's safety is forbidden, both in peace and war times.[27]

The prophet considers futile destruction to be sinful. Abdullah Ibn Umar reported that the Prophet cursed one who needlessly destroys the life of a living creature as a pastime. The Prophet (pbuh) said this when he passed by two youths from the Quraysh tribe who had tied a bird for a shooting target. In another *Hadith*, the Prophet (pbuh) warned that anyone who kills a sparrow in vain will be taken to account for it on the Day of Judgment. He also forbade the cutting down of trees that provided valuable shelter to humans or animals in the desert. The Prophet is also reported to have said that one who (wastefully) cuts down a tree invokes upon himself punishment in hell in the hereafter.[28]

Thunder

There are many *Hadiths* about lightning and thunder. Abu Sai'd al-Khudri reported: The Messenger of Allah said, "A bridge will come and be placed across Hell." We said, "O Messenger of Allah, what is the bridge?" The Prophet said, "It is shaky and slippery, over which are claws, pincers, and broad spikes with curved thorns like the seed found in Najd known as Al-Sa'dan. The believer will cross over it like the blink of an eye, or like lightning, or like wind, or like powerful horses, or like camels. Thus, one will be unharmed, another will be lacerated, and another will plunge into the fires of Hell, until the last one will be dragged across it."[29]

Prophet Mohammad (pbuh) further clarified the lightning or thunderbolt, "Do you not see that the thunder came down and went back up in the blink of an eye?"[30] As lightning connects to the ground from the clouds, a second stroke of lightning will return from the ground to the clouds, following the same channel as the first strike. Scientifically, it can be said that for the thunderbolt or lightening, positive charges gather on the earth's surface while electrons, or negative charges, formulate increasingly beneath the clouds. When a negative charge gets close to the ground, a positive charge, called a streamer, reaches up to meet the negative charge. The channels connect and we see the lightning strike. The downward strike brings electrons from the clouds to the ground at a very high speed, transferring a large number of electrons from the clouds to the ground, changing the ground from positive to negative charges. Meanwhile, the clouds change to positive charges. The sudden voltage results in a second thunderbolt that goes upward.

The phenomenon of a thunderbolt striking down and going up was mentioned by the Prophet to occur in a blink of an eye, and it was scientifically proven a thousand years later. Obviously, his statement was factual, which surprised modern scientists as to how an illiterate person knew about the natural phenomenon in detail. Lightning bolts shooting upwards were first discovered in 1993 due to advancements in slow motion camera recordings.

A thunderbolt is a natural phenomenon formed by the sudden movement of electrons from an area with dense negative electric charges to a positively charged point. This phenomenon can occur in one single cloud or two different clouds, or between a cloud and the ground. It results in a flash or lightning bolt followed by roaring thunder. If a thunderbolt hits something on the ground, its sound will bang like a large explosion. Nonetheless, nobody would think of it as two strikes happening at almost the same time.

Brain

Prophet Mohammad (pbuh) spoke of many things that were later proven to be academically correct. He once talked about the forebrain, which was later found by scientists to be related to behaviour. The Holy Quran (96:15-16) says, Yes, if he desisted not, We will assuredly drag him by catching his forelock hairs. Forelock of what type, lying, sinful."[31] Holy Quran (11:56) again says "I relied on Allah, my Lord and your Lord. There is no moving creature whose forelock (forebrain) is not under His control. No doubt! My Lord meets on a straight path."[32] In the past, scientists did not understand how the brain works, and ancient people did not even know what the brain does in general. However, the Quran describes the frontal lobe as being involved in directing decision-making on both good and bad things. To be more specific, God emphasises in the Holy Quran that he will control this particular part of the brain, which is related to wrong conduct and evil deeds.

After the verses about the forelock were revealed, Prophet Mohammad made a supplication to Allah. According to a *Hadith* reported by Imam Ahmad bin Masud, Prophet Mohammad (pbuh) said, "O Allah, I am Your slave, the son of Your slave. My forelock is in Your hand. Your judgement of me is inescapable. Your trial of me is just. I am invoking You by all the names that You call Yourself that You have taught to anyone in Your creation, that You have mentioned in Your Book, or that You have kept unknown. Let the Quran be the delight of my heart, the light of my chest, the remover of my sadness and the pacifier of my worries."[33]

Considering this report, the Prophet understood the importance of the forelock and surrendered it to Allah to direct him to the right path. The embryologist added that no scholar was aware that the forelock was related to human and animal behaviour until centuries later. No civilization appeared to even bring it up before. But the Quran was the only book that mentioned it, and the final Prophet of Islam emphasised it. This also points out that Allah gives freedom of behaviour to humans and animals. But we have been created and given sublimity, and we humans need to learn to control our behaviour, the Prophet said.

Heart

On many occasions, Prophet Mohammad (pbuh) predicted a number of events to take place before the end of the world. Among those, there is one related to science, although minor. The prophecy indicated that "when the end of the world draws near, humans will increasingly die sudden deaths."[34] This *Hadith* was reported by Al-Tabarani. However, the Prophet did not specify what would cause sudden deaths, whether it was diseases or incidents. Modern Islamic scholars have witnessed numerous deaths from heart attacks and brain deaths caused by cardiovascular diseases and believe that these lead to sudden deaths as foretold by Prophet Mohammad (pbuh).

Although the Prophet had not clarified the name of the disease, cardiovascular disease, or heart attack, are the most common causes of sudden death these days. Besides cardiovascular disease, there are reports of *Hadith* pointing out other illnesses. In the *Hadith* collection of Imam Ahmad Ibn Hanbal, it was narrated that Usamah Ibn Shuraik said, "I was with the Prophet when the Bedouins came to him and said, 'O Messenger' of Allah, should we seek medicine?" He said, "Yes O slaves of Allah, seek medicine, for Allah has not created a disease except that 'he has also created its cure, except for one illness.' They said, 'And what is that?' He said, 'Old age'."[35]

This warning conforms to modern knowledge of science and encourages scientists to search for a cure for every disease. Medical scientists admit that senility is the best mechanism to stop life. After a long period of time, life is spent; cells and tissues get weakened and stop working eventually. Numerous attempts to stop this natural mechanism result in many side effects, such as cancer and other diseases. Professor Lee M. Silver, a molecular biology expert at Princeton University in the United States, said that scientists' efforts to create eternal life have brought about the exact opposite results. One of them is cancer, which is one of the top three causes of natural death in the world. A large number of people around the globe die from it each year. Seeking immortality is medically impossible, and eventually every life has to end.[36]

About Flies

Prophet Mohammad (pbuh) said: "If a fly falls into the drink of any one of you, he should dunk it all the way in and then remove it, because on one of its wings is disease and on the other is its cure."[37] In modern times it was discovered that the common fly carried parasitic pathogens for many diseases, including malaria, typhoid fever, cholera, and others. However, after 1300 years in modern times, it was also discovered by scientists that the common fly carried parasitic bacteriophagic fungi capable of fighting the germs of all these diseases.

There are treatment properties found in flies. This thesis was based on reports from the First World War, which lasted from 1914 to 1918. According to the reports, the largest cause of soldiers' deaths was the dirtiness of wounds, which led to the dissipation of poison created by bacteria in the blood flow. There was no antibiotic or antibacterial drug at that time, and the first disinfectant was introduced in 1928. Thus, without any antibiotics, a few other methods were used to treat wounds. One of them was allowing flies to lay eggs in the wounds. Even today, fly larvae, or maggots, are used medicinally to clean up festering wounds. They only eat dead tissue and leave healthy tissue alone.[38]

It is established that house flies are carriers of dangerous pathogens for animals and humans. Even the muscaphobic critics of this *Hadith* are forced to admit that no one at the time of the Prophet knew that flies carried such harmful organisms. Whence the observation that "under one of its wings there is venom"? Second, from the perspective of logic, if the fly did not carry some sort of protection in the form of an antidote or immunity, it would perish from its own poisonous burden and there would be no fly left in the world. So, it has been now proven by modern scientists that flies carry the disease (pathogens) plus the antidote for those organisms. These scientists prove the Tradition of the Prophet that when a fly touches a liquid food it infects the liquid with its pathogens, so it must be dipped in order to release the antidote for those pathogens to act as a counter-balance to the pathogens.

Plague

At the young age of Prophet Mohammad (pbuh) and even before, a devastating pandemic known as the plague was killing the population of most of the world. Even the Emperor Justinian of the Byzantine Empire was killed due to the plague. The Prophet repeatedly warned his followers against the threat of the plague.

Seven years after the Prophet's death and under the succession of Umar Ibn Khattab, the Caliph led his troops to contain the situation at the Syrian border, where he found a serious outbreak of the epidemic. He immediately decided to pull out his forces. During his withdrawal from the region, he met Abu Ubaidah Ibn al-Jarrah, commander of a large section of his army, who was marching to support Caliph Umar. The commander asked the Caliph, why did he disobey Allah's will by calling back the troops? Umar responded, "I had followed the will of Allah by withdrawing my army from the epidemic as the Prophet warned before."[39] About the epidemic, the Prophet had already warned, "If you hear of an outbreak of plague in a land, do not enter it; but if the plague breaks out in a place while you are in it, do not leave that place."[40]

Regardless of the Caliph's warning, Ubaidah insisted on his march. As a result, up to 25,000 people, including the commander himself died by the cause of epidemic.[41] It was the first and most serious epidemic in Muslim history, and it served as an important lesson for the Islamic Empire to place a greater emphasis on water management and cleanliness in order to prevent the disease from returning.

NOTES

1. Some of the *Hadith* were memorised by only one follower of the Prophet, and to make things worse, he moved to somewhere distant. If any other followers wanted to learn the *Hadith*, they had to travel for months to meet and learn from the only one who had memorised them. A good example of the travels of the companions was that of Jabir Ibn Abdullah. He lived in Madinah and it took more than a month, covering 1,400 kilometres, to travel to Sham (present-day Syria) to see Abdullah Ibn Unais only to hear one *Hadith* that was remembered by him alone. Also, another companion named Abu Ayyub Al-Ansari travelled over 2,000 kilometres from Madinah to Egypt to hear just one *Hadith*, which was memorized by none other than Uqbah Ibn Amir. After hearing it, it took him months to return to Madinah in order to record and propagate the *Hadith*. Dahlan, Winai, *op. cit.*, p. 73.
2. *Al-Tirmidhi,* 422.
3. Nasr, Seyyed Hossein (1976). *Islamic Science: An Illustrated Study*, London, p. 4. (Muslims accept Allah not just as the only Creator or the

First Cause or the Prime Mover or the Supreme Being, but also as the only Object of Worship, Master, Sustainer, Protector, Provider, Ruler, Judge, and Ultimate Source of knowledge, wisdom, and values, for all human beings and all that exists. The essence of this absolutely monotheistic worldview is to be found in the brief Islamic testimony of faith (*Kalimat al-Shahadah*) that 'There is no object of worship whatsoever other than the One True God, Allah, and that Mohammad (pbuh) is the Messenger of Allah'.)

4. The Islamic propagation launched by the Prophet took thirteen years in Makkah, with undesirable effects. Less than 200 people embraced Islam, 100 of whom fled from escalating persecution in Makkah to the southern Kingdom of Abyssinia. The rest followed the Prophet's migration to the northern city of Yathrib or Madinah. Muslims who migrated from Makkah to Madinah were known as *Muhajirun* or migrants, and the local Muslims in Madinah were called *Ansar* or helpers who significantly increased in number and became an important support for the Islamic expansion.
5. Ameer Ali, Syed (1922). *Spirit of Islam: A History of the Evolution and Ideals of Islam with a Life of the Prophet*, London. p. 361. The Prophet of Islam himself advised his companion to spread to the whole world as far as China in the urge to acquire education and knowledge. Referring to a country like China was certainly meant to seek knowledge of scientific advancement and other worldly affairs.
6. *Holy Quran*, 13:28.
7. Ameer Ali, *op. cit.,* p. 360.
8. *Ibid.*, pp. 360-61. (Quoted from the *Bihar-ul-Anwar* of Mulla Bakir Ibn Mohammed Taki al-Majlisi, Vol. I. Chap. on Knowledge).
9. Bhargava, *op. cit.*, p. 28.
10. Ameer Ali, *op. cit.,* p. 362.
11. *Al-Tirmidhi,* Number 422.
12. *Sunan Musnad Ahòmad*, Number 16957. Surprisingly, countries with the highest growth rate of the Muslim population turn out to be those in the America, including Argentina and Brazil, where there were hardly any Muslim before. Nowadays, the two countries house millions of Muslims. In a North American country such as Canada, the Muslim population has increased from a hundred thousand to nearly 2 million in the last 20 years, whereas the United States has seen a rapid growth rate in the number of Muslims, reaching over 10 million people. Japan also used to have very few Muslims, but currently, Islamic activities are becoming widespread. These examples prove that what Prophet Mohammad (pbuh) said 1400 years ago was nothing but the truth. Dahlan, Winai, *op. cit.*, pp. 85-86.
13. *Ibid.*, p. 89.
14. *Diroyah: Jurnal Studi Ilmu Hadis* 5, 1 (2020): 21-30.
15. Dahlan, Winai, *op. cit.,* p. 88.
16. *Sunan an Nasai*, Vol. 1, Book 8, *Hadith* 737.
17. *Mishkat al-Masabih*, Book 3, *Hadith* 191.

18. Dahlan, Winai, *op. cit.,* p. 92.
19. *Ibid.*, p. 93.
20. Iqbal, Muzaffar, *op. cit.,* pp. 67-68.
21. *Ibid.*
22. *Sahih Al-Bukhari* 52, and *Sahih Muslim* 1599.
23. *Sunan Ibn Majah,* Number 3436.
24. *Sahih Muslim* 223.
25. *Sunan Ibn Majah* 274.
26. Kamali, Mohammad Hashim (2016). *Islamic Perspectives on Science and Technology*, Singapore, p. 181.
27. *Ibid.,* p. 184. The Islamic tradition is expressive of a great deal of concern for animal welfare. There are *Hadiths* on record where the Prophet warned, in particularly emphatic terms with Allah's wrath and punishment, those who are cruel to animals, such as dogs, cats, camels and cattle, under their care.
28. *Ibid.,* p. 185. The early Traditionist Abu Dawud recorded *Hadith* reports to the effect that in some parts of Madinah, the Prophet had strictly prohibited the cutting of trees and hunting of animals. These last are also prohibited during war, on the authority of *Hadith*, and established precedent of the four first Caliphs, unless it be for a manifest need or benefit. In addition to their nutritional and medicinal value for humans and animals, plants enrich the soil and protect it from erosion by wind and water; they conserve the water by hindering its run-off, moderate the climate and produce the oxygen we breathe. The Quran also mentions the aesthetic values of plants and animals that bring joy and peace of mind.
29. *Sahih al-Bukhari*, Number 1338, *Sahih Muslim*, Number 2870.
30. *Sahih Muslim*, Vol. 1, Book 1, *Hadith* 380.
31. *Al-Quran.*
32. *Ibid.*
33. *Sunan Musnad Ahmad*, No. 1/391.
34. *Sahih al-Jami*, Number 2/1026.
35. *Abu Daud,* Tibb 1, 3855; *Tirmidhi,* Tibb 2, 2039; *Ibn Majah,* Tibb 1, 3436.
36. Winai Dahlan (2015). *Muhammad: The World's Great Scientist*, Bangkok, p. 111.
37. *Sahih al-Bukhari, Number* 3320, 5782.
38. Winai Dahlan (2015). *Muhammad: The World's Great Scientist*, Bangkok, p. 119.
39. *Sahih al-Bukhari, Number* 5729.
40. *Sahih al-Bukhari, Number* 5728.
41. Dahlan, Winai, *op. cit.,* p. 187.

4
Education and Scientific Development under Muslim Rulers

Prophet Mohammad (pbuh) unified Arabia in his twenty-three years of prophethood. The four Caliphs who followed, Abu Bakr, 'Umar, 'Uthman and 'Ali, called 'Rightly-Guided Caliphs' (*Khulafa' Rashidun*), consolidated the newly founded Islamic Empire (*Dar al-Islam*) and spread its boundaries to Central Asia on the one hand and North Africa on the other. The later Caliphs of the Ummaiyad and Abbasid dynasties further expanded the Islamic Empire. A new spring of thought flooded the Arabian Peninsula and swiftly spread to more or less the whole world in the following centuries. The Muslims travelled to all corners of the world in their quest for knowledge and took up the study of natural phenomena with religious fervour and commitment, observed nature carefully and undertook extensive experimentation to gain insight and understanding of the nature of things.

Scholars broadly agree that for much of the mediaeval period, Islamic societies led the world in both technology and science. The Muslim rulers and Muslim intelligentsia gratefully and freely accepted, assimilated, integrated, and advanced the available accumulated scientific knowledge, wisdom, and learning of the existing civilizations at that time. From the seventh century on, while the rest of the world languished and experienced the Dark Age, thousands of social, scientific, and technological achievements were spawned by men and women of different faiths and cultures who lived in Muslim civilization.

The emergence of the scientific tradition in this multi-religious, multi-ethnic atmosphere was a dynamic process involving interactions between patrons of learning, scholars, scientists, rulers, guilds, and wealthy merchants. To be sure, scientific activity at this early time was not yet an institutionalised effort, but we do know that groups of scientists were already working together in the second half of the eighth century in Bagdad and other cities of the Abbasid Caliphate. It was definitely Islam which goaded the people for all-round development, and the Holy Quran and teachings of the Prophet were the torch bearers in their efforts.

It was in the field of natural sciences that Muslim scientists made outstanding advances and achieved the greatest triumph. Islam initiated in every field of human life a scientific and knowing spirit, which had never been known in the history of mankind. This upsurge of scientific activity among the early Muslims is irrefutable proof of the inherently rational and empirical orientation of the Islamic ideology and world-view. Volumes of history books and thousands of *Al-Makhtutat* (old handwritten books) tell us that Muslims contributed to the advancement of the world and the establishment of numerous branches of science.

The Muslims followed the Islamic guidelines and sincerely began to think about educational development. In spite of the upheaval of the Arab race under the early Caliphs, literature and the arts were by no means neglected in the metropolis of primitive Islam. Caliph Ali and his cousin Ibn Abbas, gave public lectures on poetry, grammar, history, and mathematics; others taught the art of recitation or elocution; whilst some gave lessons in calligraphy (in ancient times an invaluable branch of knowledge). Ali had devoted himself to the study of the Prophet's precepts by the light of reason. The Prophet himself said, "I am the city of knowledge, Ali is its door. Whoever desires to enter the city should enter from the door."[1]

Islamic Educational System

The process of learning the teachings, which were jointly recorded, memorized, discussed, and consulted, and of decision-making based on the leader throughout the 23 years of the Prophet's hard work, completely changed and softened the hearts of the Arab tribes, inducing them to acknowledge and honour others' knowledge and experiences. In Arabia during the lifetime of the Prophet, the Islamic education was introduced to his disciples by the Prophet through the process of *Da'wah* (call to Islam), it was as though a whole people went to school to read, write, and memorise their first primer, Al-Quran. Among the celebrated teachers of the Quran in early Islam were Ubadah Ibn al-Samit, Mus'ab Ibn Umayr, Mu'adh Ibn Jabal, Amr Ibn Hazm, and Tamim al-Dari. These teachers were sent to various parts of Arabia and beyond.

Islamic education begins with the lessons of the Quran. Every Muslim has a religious duty and obligation to preach and teach their fellow Muslims and non-Muslim acquaintances about the Quran and the Traditions. Such a process of informal mass education and Islamization began in the Arabian Peninsula during the Prophet's last years, and the process was carried forward under his successors. To the Muslims, the educational system was, of course, based on the traditional Islamic concept of knowledge and learning. Islam considers knowledge (*Ilm*) as something sacred because, ultimately, all knowledge concerns some aspect of God's theophanies. The Prophet told his disciples: "the word of wisdom (*Hikmah*) is the lost property of the believers. Wherever he finds it, he is most deserving of it."[2] Thus, the Prophet created an incentive to pursue all kinds of knowledge, including science and philosophy.

It is this sacred view of knowledge that has imbued the whole Islamic educational system, making it even more institutionally inseparable from specific religious organisations and institutions such as the mosque, the Sufi centre, and places run by means of endowments (*Awqaf*). This view has made the relationship between the teacher and the student in traditional schools the most intimate and spiritual one, and teachers have enjoyed the reverence. The system proved the famous words of Ali Ibn Abi Talib that "I have become the slave of him who has taught me a single word." The whole affair of education has always been at the heart of Islamic civilization as one of its basic pillars because it has been inseparable from the tradition itself, which forms the marrow and the backbone of the whole of Islamic civilization.[3]

In the whole educational system of Muslim world the imparting education was a sacred one and the teachers were most honoured. The cultivation of the Islamic sciences depended upon a vast educational system which embraced both formal and informal education and which made possible the encouragement and transmission of knowledge in all its forms. It emphasized most of all the religious sciences but included nearly all other forms of knowledge from theodicy to pharmacology.

The centres in which the Islamic sciences have been taught over the ages have been an integral aspect of Islamic civilization. From the beginning, the mosque was at once the religious and social centre of the Islamic community as well as a centre for learning. But historically, there gradually developed a distinct institution called a Madrasah (literally, the place for lessons) which grew alongside the mosque. Then, as teaching became formalised and extended, buildings were often erected specifically for teaching purposes, with mosques attached to them. Some of the greatest Muslim centres of learning, such as Al-Azhar, are even now both mosques and schools, reflecting the profoundly religious character of learning in

Islam. The Madrasah developed into a full-fledged university around eleven hundred years ago, a university in which a variety of subjects, from religious law to astronomy, were taught regularly.[4]

Some of the earliest Islamic educational institutions dating from this period of genesis still survive, the best examples being perhaps the Qarawiyyin in Fez, the Zaytuniyyah in Tunis and Al-Azhar in Cairo. Other outstanding Madrasahs, such as those of Qum, Mashhad, Samarqand, Isfahan, Najaf etc., and the like, belong to later periods of Islamic history, but are profoundly connected to the earlier Madrasahs. Likewise the mediaeval European universities and their curricula reflect their close relation with the Muslim universities, which they emulated consciously. Apart from the religious learning many of the intellectual sciences such as philosophy, logic, and mathematics have been taught in various Madrasahs over the ages and are still taught today.[5]

Apart from the education of religious sciences and intellectual sciences taught in the Madrasah, mosques, and Sufi centres were also used for the same purpose. Not only are initiatic and spiritual practices performed in Sufi centres, but many esoteric and even exoteric sciences are also taught. The Sufi centres became especially important as centres of formal learning after the Mongol invasion, when they were forced to fulfil the role of the Madrasahs destroyed by the Mongol onslaught as well as their own proper function of training the adepts' spiritually.[6]

There are also private circles connected with a particular teacher, meeting either at his house or the house of one of his disciples. Much of traditional learning is carried out today in the Islamic world in such private circles, whose great propaedeutic importance can hardly be overestimated. Moreover, the occult sciences such as alchemy or geomancy have always been taught in completely private circumstances. The atelier of an alchemist is almost always his home or the house of an adept, and it is here that those considered by the alchemist as capable of mastering the royal art have been and are to this day trained.[7]

Education has always ranked highly in Islamic society during the Middle Ages. During this age in the Muslim world, education centres soon became widely spread. There was no village without a mosque, an elementary or secondary school for teaching. They were supported by the endowments granted to them. It was mainly due to the teachings of the Holy Quran that urged Muslims to learn the principles of various branches of science. As for educational development under some of the Umayyad and most of the Abbasid rulers, the educational system gained momentum and many universities were established. The Muslim universities and the educational system in the Middle East appeared long before similar establishments in Europe. It was familiar and common practice among the Europeans to go and study in Muslim universities, notably in Spain under Muslim rule, where there was no hindrance to scientific research.

In mediaeval times and seven centuries of Muslim rule in Spain the Europeans gained, among other things, access to the accumulated treasures of Greek and Islamic learning. But, on the other hand, the protracted and bitter confrontation during the Crusades, and later the Ottoman domination of the Balkans, left on both sides a heritage of prejudice and resentment. This feeling of hostility caused the differences between the two civilizations to be magnified.

Under the Umayyads, the Muslims passed through a period of probation, preparing themselves for the great task they were called upon to undertake. Under the Abbasids, we find the repositories of the knowledge of the world. Every part of the globe was ransacked by the agents of the Caliphs for the hoarded wealth of antiquity; these were brought to the capital, and laid before an admiring and appreciating public. Schools and academies sprung up in every direction; public libraries were established in every city, free to every citizen; the great philosophers of the ancient world were studied side by side with the Holy Quran.

Needless to say that Muslims set ideals of civic-life, family life, hygiene, agriculture, architecture, irrigation, music, dress, food, calligraphy and games for every country they came into contact with it. They dived deep into the ocean to bring out pearls and penetrated low into the earth's crust to dig out ores of iron, silver, gold, lead, mercury, and the like. The Muslim scholars, built up on the teachings and fundamentals laid down by the Holy Quran, achieved basic creative contributions of their own in various branches of science, notably mathematics, astronomy, meteorology, physics, and medicine. During the Abbasid period, paper manufacturing became an indigenous industry. Books were thus written and sold on a large scale by book agencies and libraries.

With the continuous development of the Islamic sciences, new branches and forms of science came into being, and at the same time, the sciences which were taken over from the pre-Islamic civilizations were ever more Muslimized and fitted into the Islamic hierarchy of knowledge. The Muslim scientists classified the various branches of science which were studied and taught at various Islamic institutions and research centres. Al-Kindi, Abu Nasr al-Farabi etc. wielded a much wider influence over the curricula of Muslim universities and even those of the West. They gave order to the sciences and classified them.

Umayyad Caliphate (661-750)

The first four Caliphs—Abu Bakr, Umar, Othman, and finally Ali—were later referred to as the Rightly-Guided Caliphs, and each Caliph was chosen anew rather than having the Caliphate passed down to their children. However, Abu Bakr was the only one of them who did not meet with a violent end. It was probably to stabilise that situation that a powerful Makkan family called the Umayyads, led by Muawiya, eventually took control after Caliph Ali was murdered in 661. Umayyad rule started with the accession of Amir Muawiya Ibn Abi Sufyan to the throne of Caliph at Damascus in 661 CE. He was the first to change his capital from Makkah to Damascus.

During the Umayyad rule, the Caliphate became less 'apostolic' and more 'political' in character, although still of a completely religious nature, there being no secularism in Islam. Umayyad rule was divided between two branches of the family; the Sufyanids (661-684)—the descendants of Abu Sufyan; and the Marwanids (684-750)—the descendants of Marwan I Ibn Al-Hakam. They changed the old democratic setup and Amir Muawiya, the first ruler of this dynasty, secured during his lifetime an oath of allegiance to his son Yazid I, who disregarded the traditional election (*Bayaah*) and introduced the alien concept of hereditary succession. Under Umayyad rule, the Muslim armies laid the foundations for expansion of the empire as far as Sind, Samarqand, and Tashkent. The Umayyad rule began to decline when the revolt started against them in Syria, Iraq, and Khorasan in 745-750.

From the rise of Islam in the seventh century until the start of the Abbasid Caliphate in 750, the nascent Islamic world produced relatively little scientific output. At the time of the Umayyad rule, the acquisition of knowledge of science and learning was dead slow, and in other words, a little blows to the progress of knowledge and liberalism in the Muslim world. During the whole period of the Umayyad rule, the Arabs constituted the dominant element, the aristocratic military caste amongst their subjects. The majority of them were occupied in warlike pursuits. The whole Umayyad period was full of bloodshed and tyranny. The Episode of Karbala was held. Makkah was attacked and the Holy Kaabah was burnt.

The Umayyad produced only one man devoted to the cultivation of letters, and this man was Abu Hashim Khalid Ibn Yazid, the philosopher of the Marwanid family. Another Caliph, Umar Ibn Abdul Aziz or Umar II, tried to infuse religious zeal into the administration. Umar II attempted to revert to the practice of Umar Ibn Khattab or Umar I. He also abolished the infamous practice of the public cursing of Caliph Ali in the mosques, and thereby gained the sympathy of all Muslims, including the Shias.[8]

Abbasid Caliphate (750-1258-1517)

A revolt started against Umayyad rule in Syria, Iraq, and Khorasan in favour of the Abbasid claim to the Caliphate, which was moving westward under the leadership of Abu Muslim, who had captured the city of Kufa by the middle of 749. Early in 750, Abu'l Abbas (posthumously called Al-Saffah) was proclaimed the first Abbasid Caliph at Kufa. Abu Al-Abbas Al-Saffah (750-754) thereby became the first Caliph of the Abbasid dynasty. Two months later, the Umayyads were decisively defeated at the battle of Greater Zab, and by June 750 most of them had been massacred. Abd Al-Rahman I was the sole survivor from the ruling family; he escaped to Spain, where he established Umayyad rule (755-1031). The Abbasid dynasty was descended from Prophet Mohammad's (pbuh) uncle Abbas Ibn Abdul Muttalib (566-653) from whom the dynasty took its name. The Abbasid Caliphate centred its government at Kufa at first, but after Caliph Al-Mansur founded Bagdad in 762, the capital was changed to Bagdad.[9]

In the Abbasid Caliphate, Arabic was the language of court. diplomacy and administration, and the economic progress was accompanied by an outburst of intellectual activity such as Asia had never witnessed before. The Abbasid dynasty continued to rule over most of the Muslim world until the siege of Bagdad by the Mongol conqueror Hulegu in 1258. Before that, in 945, Ahmad Ibn Buya of the Buyid dynasty conquered several territories and was officially recognised as the Amir of Iraq. Although he was an Amir under Caliph but was politically more powerful and also made his capital in Bagdad. From this date, the Caliphs merely held a court, but governed no empire, until their extinction by the Mongols in 1258. Even after Hulegu's assault, the Abbasid Caliphate continued but remained only a spiritual power under the Seljuks Sultans until 1517.[10]

Pakistani physicist Pervez Hoodbhoy has rightly described the golden days under the Abbasid Caliphate. He argued that a thousand years ago there was an Islamic golden age of reason and science under the Abbasids, a period in which theology was dominated by the rationalist Mutazilites and science and philosophy by translations of Greek works. This age of tolerance and creativity, Hoodbhoy claimed, came to an abrupt end when Imam Ghizzali attacked logic and science in the name of an anti-rationalist Ash'arite theology. Thereafter, the Islamic world settled into a dogmatic slumber that has not yet ended, as evidenced by the miserable state of science in the Islamic world.[11]

The Abbasid Caliphs sponsored a translation movement aimed at rendering every available scientific text into Arabic. This translation movement coincided with and served as a catalyst for the explosion of scientific output that occurred in the Islamic world over the following centuries. Scientists during this period made important advances in fields as varied as astronomy, mathematics, medicine, and optics. Rulers and other wealthy individuals patronised scientists for both prestige and for the practical benefits promised by the practitioners of medicine, astronomy, astrology, and applied mathematics.

Abu Ja'far Al-Mansur (754-775)

A new era of educational and scientific achievements began with the rise of the Abbasids. They rose to power with the assistance of the Persians, and they relied for the maintenance of their rule upon the attachment of the general body of their subjects. After Al-Saffah, his brother Abu Ja'far Al-Mansur (754-75) became the second Caliph of the Abbasid dynasty. Al-Mansur was a statesman of the first rank, and his method of government gave a new bent to his genius. They settled in cities, acquired landed properties, and devoted themselves to the cultivation of letters with the same order which they had displayed in the pursuit of war.

Caliph Al-Mansur founded Bagdad city as his capital on the bank of the river Tigris. The new capital was planned as a round or circular city with sixteen gates. The city was officially called Madinat Al-Salam, or the City of Peace. It was built on the site of an ancient village, planned by four eminent architects, and built by 1,00,000 workers and craftsmen over a period of four years (762 to 766). Its construction began on 30th July 762, a date determined by astrologers and engineers and it remained the intellectual capital of the world for five centuries. In the days of its glory, before the destroying hordes of Hulegu Khan, Bagdad presented a beautiful and imposing appearance, a fit capital for the Pontiffs of Islam. Bagdad had become the intellectual and scientific capital of the Abbasid Empire, providing scientists with patronage and opportunities to experiment, discuss, and discover.[12]

The great number of holy men, who have found their last resting-place within or about its walls, and whose tombs are objects of veneration to all Muslims, gave to Bagdad the title of Bulwark of the Holy. Here are the mausoleums of the greatest Imams and the most pious Sheikhs. Here reposes the Imam Musa Al-Kazim, and here are buried Abu Hanifa, Shaikhs Junaid, Shibli, and Abdul Kadir Ghilani, the chiefs of the Sufis. The monuments of the Caliphs and their consorts stood alongside those of the Imams and Shaikhs.[13] Many of the academies, colleges and schools were filled in the Bagdad city. Two institutions surpassed all others in importance by their wealth and the number of their students. These were the Nizamiah and Mustansarieh; the first was established in the first half of the fifth century Hijri by Nizam ul-Mulk, the great Vazir of Malik Shah, Sultan of the Seljuks; and the second was built two centuries later by the Caliph Al-Mustansir Billah.[14]

It was at Caliph Mansur's command that literary and scientific works in foreign languages were first translated into Arabic. He himself was no mean scholar and mathematician. He had the famous collections of Indian fables (*Hitopadesa*), the Indian treatise on astronomy called the *Siddhanta*, several works of Aristotle, the *Almagest* of Claudius Ptolemy, and the books of Euclid, as well as other ancient Greek, Byzantine, Persian, and Syrian productions, translated into the language of the Arabs. Masudi mentions that no sooner were these translations published than they were studied with much avidity.[15]

Mansur's successors were not only warm patrons of the learned, who flocked to the metropolis from all quarters, but were themselves assiduous cultivators of every branch of knowledge. Under them, the intellectual development of the Saracens, in other words, of the conglomerate races of the vast empire which constituted the Caliphate, proceeded with wonderful rapidity. Under the first six Abbasid Caliphs, but especially under Caliph Al-Mamun, the Muslims formed the vanguard of civilization. Under the inspiring influences of the great Prophet, who gave them a code and a nationality, and assisted by their sovereigns, the Saracens caught up the lessons of wisdom from the East and the West, combined them with the teachings of the Master, and "started from soldiers into scholars."[16]

Al-Mamun (813-833)

The most important Caliph after Al-Mansur was seventh Caliph Al-Mamun (813-833). He was known for his attempt to end the sectarian rivalry in Islam and to impose upon his subjects a rationalist Muslim creed. Well educated and with a considerable interest in scholarship, Al-Mamun promoted the Translation Movement, the flowering of learning and the sciences in Bagdad and the publishing of Al-Khwarizmi's book Algebra. Under Mamun, the *Almagest* of Ptolemy was re-translated, and the verified Tables prepared by famous astronomers like Sanad Ibn Ali, Yahya Ibn Abi-Mansur, and Khalid Ibn Abdul Malik. Their observations connected with the equinoxes, the eclipses, the apparitions of the comets, and other celestial phenomena, were valuable in the extreme, and added greatly to human knowledge.[17]

In 820 or the early ninth century, while Europe was still in the Dark Ages, believing that the globe was a flat surface surrounded by oceans, Caliph Al-Mamun of Bagdad ordered scholars to split into two teams with Sanad Ibn Ali and Ali Ibn Isa al-Asturlabi as the leaders to measure the world's perimeter. As a result, they not only concluded the length of the equator to be 39,986 kilometres (which deviated 89 kilometres or 0.22 per cent from the correct number), but they also determined the world's longitude at 360 for the first time. Longitude is the geographic coordinate that determines the difference in time based on the earth's rotation. This points out that the Muslim world discovered that the earth was round a thousand years ago, well before the Europeans included this knowledge in their school lessons.[18]

By the orders of Al-Mamun, Mohammed Ibn Musa Al-Khwarizmi made a new translation of the *Siddhanta*, or the Indian tables, with notes and observations. Al-Kindi wrote two hundred works on various subjects: arithmetic, geometry, philosophy, meteorology, optics, and medicine. He was well-versed in Greek, and he obtained some of the knowledge he embodied in his invaluable treatises from the schools of Athens and Alexandria. Abd-Ma'shar (Albumazar) made the celestial phenomena his special study; and the Zij-abi-Ma'shar, or the Table of Abu-Ma'shar, has always remained one of the chief sources of astronomical knowledge. The discoveries of the sons of Musa Ibn Shakir, who flourished under Mamun and his two immediate successors, especially with respect to the evaluations of the mean movement of the sun and other astral bodies, are almost as exact as the latest discoveries in Europe.[19]

These scientists ascertained with wonderful precision, considering the appliances they possessed, the obliquity of the ecliptic and marked for the first time the variations in the lunar altitudes. They also observed and determined with remarkable accuracy the precession of the equinoxes and the movements of the solar apogee (which were utterly unknown to the Greeks). They calculated the size of the earth from the measurement of a degree on the shore of the Red Sea—this at a time when Christian Europe was asserting the flatness of the globe.

Abu'l Hasan invented the telescope, of which he speaks as "a tube to the extremities of which were attached diopters." These "tubes" were improved and used afterwards in the observatories of Maragha and Cairo with great success. The influential thirteenth-century Maragha Observatory in Iran was a scientific institution where astronomers challenged received wisdom about the universe and developed new mathematical models on which Renaissance scholars relied. By the time Al-Battani appeared, the Muslims had evolved from the crude astronomy of the ancients to a regular and harmonious science. He is, however, best known in the history of mathematics as the introducer of the sine and cosine instead of the chord in astronomical and trigonometrical calculations.[20] Al-Battani, though surpassed by his successors, occupies a high position among astronomers, and a competent judge pronounces his role to be the same among the Saracens as that of Ptolemy among the Greeks. His Astronomical Tables, translated into Latin, furnished the groundwork of astronomy in Europe for many centuries.

During the time of Al-Mamun, cartography, the science of the production of maps and construction of projections and designs, was also developed. The administrative needs of the newly conquered lands required detailed descriptions, and early maps emerged on the basis of first-hand information about the new regions. Cartography became a basic need of the expanding Muslim world within the first generation. Under Caliph Al-Mamun, a word map was constructed and named after him as Al-Suraht Al-Mamuniyya.[21]

Caliph Mamun was followed by a brilliant succession of princes who continued his work. Under him and his successors, the principal distinguishing feature of the school of Bagdad was a true and strongly marked scientific spirit, which dominated all its achievements. The deductive method, hitherto proudly regarded as the invention and sole monopoly of modern Europe, was perfectly understood by the Muslims. "The Arabs of the ninth century were in possession of that fecund method which was to become, long afterwards, in the hands of the moderns, the instrument of their most beautiful discoveries." Among the numerous astronomers who lived and worked in Bagdad at the close of the tenth century, the names of two men, Ali Ibn Amajur and Abu'l Hasan Ali Ibn Amajur, generally known as Band-Amajur, stand prominently forward. They are noted for their calculation of the lunar movements.[22]

The philosophers like Galen (Jalinus), Dioscorides, Themistius, Aristotle, Plato, Euclid, Ptolemy, and Apollonius received their due appreciation during the Caliphate period. The sovereigns themselves assist at literary meetings and philosophical disquisitions. For the first time in the history of humanity, a religious and autocratic government is observed to ally itself with philosophy, preparing and participating in its triumphs. Every city in the Islamic empire sought to outrival the other in the cultivation of the arts and sciences, and governors and provincial chiefs tried to emulate the sovereign. Travelling in search of knowledge was a pious duty. From every part of the globe, students and scholars flocked to Cordova, to Bagdad, and to Cairo to listen to the words of the Muslim sages. Even Christians from remote corners of Europe attended Muslim colleges. The men who became in after-life the heads of the Christian Church (Gerbert studied in Cordova, afterwards became Pope Sylvester II), acquired their scholarship from Islamic teachers.[23]

From the shores of the Atlantic eastward to the Indian Ocean, far away even to the Pacific, resounded the voice of philosophy and learning, under Muslim guidance and inspiration. Under the Abbasid rulers, ample consideration was given to medicine and public health precautions. In the Islamic world, during the eighth century, the first hospital was built in Damascus, having separate wards for males and females, and special wards for internal diseases, surgery, orthopedics, and other diseases. These hospitals were to become models for hospitals as we know them today. At the same time another hospital was constructed in Bagdad in 766. The Muslims also used camel convoys as transportable hospitals, which stimulated from place to place. Since the religion did not prohibit it, Muslim scholars used human bodies to study anatomy and physiology and to support their students' realisation of how the body works. This pragmatic study allowed surgery to mature very quickly.

By the end of the tenth century, the Caliphate had lost all power and vigour. What had been once a real unity under a Caliph was now a collection of scattered dynasties, not one of which obeyed the Caliph's mandate. Spain and Africa, including Egypt, had become independent of the Caliphs of Bagdad; northern Syria and arid Mesopotamia were in the hands of Arab chiefs, some of whom founded their dynasties; Persia was split up into numerous factions of the Buwayid princes, who, being Shias, had no respect for the Bagdad Caliphs. Soon, however, the whole of the northern part of the African continent was brought under the domination of the Bani-Fatima, and then another era of glory for arts and literature commenced.[24]

This glorious period of the House of Abbas lasted until the fall of Bagdad on 10 February, 1258, by Hulegu, the grandson of Genghis. For three days, the Tartar chief gave up the town to rapine and slaughter. On the third day, the thirty-seventh Abbasid Caliph Mustasin was put to death, thus ending the Abbasid dynasty.[25] Some of the Abbasid relatives took refuge in Egypt. The Mamluk Sultans searched for the Abbasid relatives who had taken refuge in Cairo and recognised one of them as a Caliph. The Abbasid Caliphs were only titular and their functions remained purely spiritual. The Islamic scientific tradition, on the other hand, had already found a new home in other Islamic lands, such as modern-day Turkey, Syria, Egypt, and Iran.

The independent principalities which were formed in Fars, Khurasan, Turkistan, Mesopotamia, Syria, Egypt, and Africa prior to the foundation of the Fatimid Caliphate in Egypt, were all in the hands of the princes who delivered the *Khutba* (Friday Sermon) in the name of the Caliph of Bagdad, and sent him fixed annual payments, though the Caliphs had lost all power and authority and were under the rule of their Turkish guards. The same system lasted till the last Abbasid Caliph resigned his claim to the Ottoman Turk Sultan Salim I after his conquest of Egypt in 1517. From this time onward, the Ottoman Turks assumed the responsibility of the Caliphate. The temporal and spiritual functions were again combined in the Caliphate.[26]

During the time of the Abbasid Caliphate, the Islamic Empire began slipping into a spiral of decline. The provincial governors and generals tried to make themselves independent and even declared their own Caliphate. Of these powers, the Fatimids, a family that claimed descent from the Prophet Mohammad (pbuh), through his daughter Fatima, declared their own Caliphate in North Africa in 909. Twenty years later, the Umayyad leader in Cordoba, Abd al-Rahman III, declared himself Caliph too. So, there were three separate Caliphs, one in Bagdad, one in Egypt, and one in Spain, and dozens of local rulers who paid them no heed.

Besides the above ruling families who separated them from the Abbasids, there were many military generals who rose in power and overpowered the Abbasids. In 945, one family of generals, the Buyids, marched south to Bagdad from the Elborz mountains in the north of Iran and stormed the city. The Buyids left the Caliph in power nominally, but adopted the titles of sultan and Shahanshah (king of kings) for themselves, reminiscent of the old Sassanian kings. In 1040, another group called the Seljuk Turks invaded Syria and Mesopotamia and, five years later, captured Bagdad. As with the Buyids, the Seljuks called themselves sultans and let the Abbasid Caliphs remain in Bagdad.

Mongol and Tartar Invasion

The Mongols[27] and Tartars[28] were the Central Asian blood-thirsty invading tribes who invaded the cultured Muslim Empire and tried to devastate them completely. In 1219, Genghis Khan and his Mongol army conquered China with swiftness and ease, and then turned westwards. With an army numbering up to 800,000, many with incomparable skill on horseback, the Mongols were unstoppable. Bukhara was quickly overwhelmed, and when they reached the university town of Nishapur, the slaughter was dreadful. The men, women, and children in the city were beheaded, then disemboweled. In other cities that fell in the Mongols' path, people were rounded up and slaughtered for sport, like cattle. Those that survived the onslaught faced years of starvation, as the Mongols destroyed the canals, the tunnels that supplied the fields with water.[29]

When Genghis withdrew, the people of Islam might have sighed with relief. But the Mongols were not gone. In 1256 and 1258, they were back, now led by Genghis's grandson Hulegu. This time, even Bagdad was not safe. Ignoring the Caliph's warning that his death would bring chaos to the world, the Mongols entered the city, killed the Caliph, and slaughtered hundreds of thousands of people. The Mongol victory was accompanied by indiscriminate killings, the destruction of all major public buildings, including shrines, mosques, Madrasahs, and palaces, and the uprooting of intellectual life.[30] No one knows how many citizens of the Islamic empire died at the hands of the Mongols, or how many died of starvation in their wake, but estimates suggest that it was many, many millions. A historian of the time suggested that the blow was so devastating that the population would take a thousand years to recover. He was not far wrong. The population of the region finally reached its pre-Mongol levels again just a few decades ago.

The slaughter of those years is still etched deep in the memories of people in these parts. Yet there was still another dreadful invasion from the east to come in 1384. This was the Tartar army of Tamerlane, which left a tower of heads piled in the square at Isfahan, and another in Bagdad, painfully rebuilt in the century after the Mongol destruction. It would be easy to think that these unimaginable traumas would have spelled the death of Islamic science and culture. However, it is not fully true because the same Hulegu Khan, the destroyer of Bagdad, became a patron of one of the greatest Muslim astronomers, Nasir al-Din al-Tusi. In Iran, when the slaughter stopped, Tamerlane and his Timurid successors presided over the continuing flowering of Iranian culture. Indeed, some of mediaeval Islam's greatest scientific achievements occurred after the Mongol and Tartar invasions.

Moorish Rule in Spain (711-1492)

In 711, a group of North African Muslims, led by the Berber general, Tariq Ibn Ziyad, landed on Jebel Tariq, or Gibraltar, to commemorate the mountain of Tariq. He fought the battle of Xeras with 12,000 Arab soldiers, while Roderic the Goth had about 100,000 soldiers. Tariq Ibn Ziyad won the battle and captured the Iberian Peninsula known as Al-Andalus (southern region of Spain). Spain was conquered by the Muslims in 710-712 and was ruled, like other provinces of the Caliphate, by governors appointed by the Umayyad Caliph until 755-756. Islamic Spain, or Al-Andalus, had a unique society that was a fusion of Christian, Jewish, and Islamic culture and an oasis of tolerance, education, and civilization. Al-Andalus quickly became the most sophisticated, intellectually advanced, and socially tolerant society in the mediaeval West.

Though most of the members of the Umayyad family had been massacred by the Abbasids, Abd al-Rahman I, a grandson of Hisham, the tenth Umayyad Caliph, managed to escape from Bagdad, and after years of wandering, he accepted the offer of a powerful section, and was welcomed on his arrival in Spain. Abd al-Rahman I landed at Mosara, was saluted as Amir on his arrival, defeated the governor, Yusuf, and was acknowledged as the Amir or Sultan of Spain in July 756. The Amir Abd al-Rahman III was the first to declare himself Caliph (separate from the Caliphate of Bagdad) at Cordova in 929.[31]

By 1000 CE, Muslims had reached the limit of their rule in Spain. Only the provinces of Spain in the extreme north, such as Leon, Burgos, and Navarre, remained unconquered. The rest of Spain was Muslim, and this included the bulk of Portugal. Under the Arabs, Andalusia or Spain, became a prosperous cultural and economic centre where education and the arts and sciences flourished. They promoted mathematics, astronomy, medicine, literature, and the fine arts.[32]

These rulers brought with them a sophisticated culture that prized tolerance and scientific learning, something of a contrast to the rest of Europe during the Middle Ages. Scholars and engineers flourished across the peninsula, especially in Al-Andalus. Here, Muslims, Christians, and Jews all lived and worked side by side. In Spain, the Moorish rule expanded scientific knowledge and inventions from the Pyrenees to the Straits: Seville, Cordova, Granada, Murcia, Toledo, and other places possessed their own public libraries and colleges, where they gave free instruction in science and letters. In the days of its splendour, Spain assumed the dignity of systematic speculation in the philosophy of the Cordova School and in the scientific research of Arab scientists.[33]

To the words of Renan, "The taste for science and literature had, by the tenth century, established, in this privileged comer of the world, a toleration of which modern times hardly offer us an example. Christians, Jews, and Muslims spoke the same tongue, sang the same songs, and participated in the same literary and scientific studies. All the barriers which separated the various peoples were effaced; all worked with one accord in the work of a common civilization. The mosques of Cordova, where the students could be counted by thousands, became the active centres of philosophical and scientific studies."[34]

The Arab writers, and their Jewish collaborators, concentrated on philosophy and the positive sciences, and achieved success. The first observatory in Europe was built by the Arabs. The Giralda, or tower of Seville, was erected under the superintendence of the great mathematician Jabir Ibn Afiah in 1190 for the observation of the heavens. Umar Ibn Khaldun, Yakub Ibn Tarik, Muslimah Al-Maghribi, and Abul Walid Mohammed Ibn Rushd were some of the physicists of Moorish Spain.[35] For nearly seven hundred years, it was the centre of Muslim culture and learning. Spain has shown extraordinary changes in the evolution of her people and the development of her civilization.

Abd al-Rahman III (912-961)

Abd al-Rahman III, the son and successor of Abd al-Rahman I, was the greatest and noblest of Muslim rulers in Spain. He abandoned the titles of Amir and Sultan and adopted the title of Caliph in 929. So he was the first to establish the Umayyad Caliphate in Spain. He was the greatest and the noblest of Muslim rulers in Spain. He not only exercised absolute sway over his subjects and kept the Christian kings of Leon, Castile, and Navarre in check, but warded off danger from Africa, and built up a powerful navy that maintained his authority in the Mediterranean. Agriculture, manufactures, trade and commerce, arts and sciences etc., greatly flourished. The traveller's eyes were gladdened on all sides by well-cultivated fields irrigated to scientific principles, so that what seemed the most sterile soil was rendered fertile. Like most great princes, he had a passion for building, and in 936, he laid, about a league to the north of Cordova, the foundations of a town designed to perpetuate the name of Zahra. Nothing was spared to make it as magnificent as possible. For twenty-five years, ten thousand workmen, provided with fifteen, hundred beasts of burden, laboured at its construction, yet it was unfinished at its founder's death.[36]

Al-Hakam II (961-976)

Al-Hakam II, the son of Abd al-Rahman III, the famous Umayyad Caliph of Spain, is well-known for his generous patronage of learning and scholarship. Under his rule, the primary schools were good and numerous. In Ardalusia, nearly everyone could read and write, while in Christian Europe, people in the most exalted positions—unless they belonged to the clergy—remained illiterate. Grammar and rhetoric were also taught in the schools. Al-Hakam founded in the capital (Cordova) twenty-seven seminaries, in which children of poor parents were educated gratuitously, the teachers being paid out of the Caliph's privy purse. Al-Hakam had constructed a magnificent library in Cordova, which housed 500,000 books. He himself had studied many of these and made brief marginal comments on them. This rich collection laid the groundwork for intellectual study in Spain and provided an excellent opportunity for later philosophers to develop their abilities. The University of Cordova was then one of the most renowned in the world.[37]

Overtime, Christians resented Moorish rule and challenged Muslim territorial dominance in Al-Andalus and slowly expanded their territory. In 1492, Catholic monarchs Ferdinand II of Aragon and Isabella I of Castile won the Granada War and completed Spain's conquest of the Iberian Peninsula. The poor Muslim ruler, Boabdil, surrendered after defeat and left Granada while weeping, leaving its fate to the Christian conquerors. Ferdinand II of Aragon and Isabella I of Castile joined their two kingdoms to create modern Spain. The Muslims had covered the land with colleges and schools; unfortunately, the Christians transformed them into churches for the worship of saints and images.[38]

The literary and scientific treasures amassed by the Muslim sovereigns were consigned to the flames. Cardinal Ximenes burned five thousand books, splendidly adorned with gold and silver and priceless illuminations, after the conquest of Granada by the Christians in 1492. The Muslim men, women, and children were ruthlessly butchered or burnt at the stake; the few who were spared were reduced to slavery. Those who fled were thrown on the shores of Africa as helpless beggars.[39]

Ali Dynasty of Morocco (785-985)

In the year 785, an insurrection of the partisans of the family of Ali took place at Madinah, and among those who took part in it was Idris, a descendant of Abu Talib. The revolt was suppressed, and Idris fled to Egypt and then to Morocco, where he founded the Ali dynasty and reigned in Ceuta. The dynasty lasted until 985 and reached its greatest extent about 860.[40]

Aghlabid Dynasty of Ifriqiyah (800-909)

The Aghlabid dynasty was founded by Ibrahim bin Aghlab in Ifriqiyah (Tunisia and eastern Algeris). The Aghlabids were vigorous rulers who employed large fleets in the Mediterranean, harried the coasts of Italy, France, Corsica, and Sardinia, and conquered Sicily in 827-28. Sicily remained in Muslim hands until its conquest by the Normans. The Aghlabid domination in Africa, when at its best, witnessed Arab supremacy in the Mediterranean. The Arab Corsairs were the terror of the Mediterranean Sea, and besides Sicily, the Arabs took Malta, Sardinia and even invaded the suburbs of Rome.[41]

Fatimid Caliphate of Egypt (909-1171)

Among those who succeeded the Abbasids, it was the Fatimids who perhaps did most, as a Caliphate, to continue advances in science and learning. The Fatimids had come to power on the basis that they had a relationship with the family of Prophet Mohammad (pbuh) and were, therefore, the legitimate heirs to the Prophet. They belonged to the Ismaili tradition, a branch of Shia Islam. The Abbasid administration was multi-cultural and multi-religious yet increasingly unpopular with Muslims, in part because of the interrogation of those who refused to agree with their rationalist take on religion. The ending of the Abbasid inquisition by the Caliph Al-Mamun weakened them further, and the close of the ninth century, therefore, saw pockets of dissent in many places, as well as, for the first time, the emergence of a power in the Fatimids, who were strong enough to challenge the Abbasids.

In 909, the Ismaili succeeded in establishing their rule in Egypt under a member of the house of Ali, Abu Mohammed Abdullah (Obaidullah Al-Mehdi). They founded the Fatimid Caliphate and ruled northern Africa for three centuries. After their conquest of Egypt, the Fatimids were the most powerful princes of Islam, and it seemed at times that the Abbasid Caliphate would disappear. The Fatimids were excellent rulers, and they made Egypt one of the most prosperous countries in Africa. The Fatimid Empire was extended from Mauritania to the confines of Egypt. The Fatimids of Egypt were grand supporters of learning and science. Fez, Miknasa, Segelmessa, Tahart, Tlemcen, Kairowan, but above all, Cairo, became centres of culture and learning.

The Fatimid rulers encouraged revolutionary movements in foreign countries. They supported the Carmathians (followers of Karmak) in Arabia, who recognised fully the temporal and spiritual authority of the Fatimid Caliphs. The Fatimid Caliphs and the Carmathians had an informal understanding and worked in secret. The Carmathians threatened the capital of the Abbasids, preyed upon the pilgrim caravans, and finally, during the pilgrim festival, forced their way on one occasion into Makkah, perpetrated a horrible massacre, and carried off the Black Stone (*Hijr-e-Aswad*) of the Kaaba (930). This dreadful act threw the Muslim world into consternation. The Fatimid Caliph, who seems to have been aware of the secret designs of the Carmathians, disavowed this act, but we know that they acted on his suggestion, and they subsequently (951), after twenty years, at the command of his successor, again restored the Holy Stone for a heavy payment.[42]

Al-Muizz li-Din-Allah (932-975)

Al-Muizz li-Din-Allah was the fourth and most important Caliph of the Fatimid dynasty in Egypt. He wrested Egypt and a portion of Syria from the Abbasids. In promoting science and knowledge, he was known among the scholars as Mamun of Egypt. To mark his victory, Al-Muizz founded Cairo (Kahira, the Victorious City), and transferred his capital from Mehdi to Cairo. At this time, his dominions included, besides the whole of northern Africa, the islands of Sardinia and Sicily.

Al-Muizz, founded in Cairo in 972, the House of Wisdom (*Dar-ul-Hikmah*), and a brilliant mosque. *Dar-ul-Hikmah* was essentially an academy for teaching astronomy, mathematics, medicine, and astrology, but it also taught the Shia beliefs of its Fatimid patrons, which occasionally inspired the resentment of the predominantly Sunni population. The *Dar-ul-Hikmah* in Cairo was just one of a few houses of wisdom that had been established throughout the Muslim world, the most famous of which was founded in Bagdad in the ninth and tenth centuries. The Fatimid rulers were patrons of letters and their colleges, public libraries, and scientific institutes were richly furnished with books, mathematical instruments, to which were attached numerous professors and attendants. At these colleges, students received instruction in grammar, literature, the interpretation of the Koran, jurisprudence, medicine, mathematics, and history.[43]

Al-Aziz Bi'llah (975-996) and Al-Hakim Bi-Amr Allah (996-1021)

Cairo had become a new intellectual and scientific centre under the Fatimids of Egypt. Here flourished, in the reigns of Aziz B'illah and Hakim Bi-Amr Allah, the master spirits of the age, Ibn Yunus, the inventor of the pendulum and the measurement of time by its oscillations. He is, however, famous for his great work named after his patron and sovereign, Zij-ul-Akbar-Al-Hakimi, which soon displaced the work of Claudius Ptolemy. It was reproduced among the Persians by the astronomer-poet Umar Khayyam (1079); among the Greeks, in the Syntax of Chrysococca;

among the Mongols by Nasir ud-din Tusi, in the Zij-ul-Khani; and among the Chinese, in the astronomy of Co-Cheou-King in 1280; and thus, what is attributed to the ancient civilization of China is only a borrowed light from the Muslims.[44]

After Ibn Yunus' death in 1009, his discoveries were carried on by Ibn un-Nabdi, who lived in Cairo in 1040, and Hasan Ibn al-Haitham, a distinguished astronomer and optician known in Europe as Alhazen and famous for discovering atmospheric refraction. Hassan Ibn al-Haitham, an experimental physicist from the eleventh century, helped to modernise our understanding of vision and is credited with describing an early imaging device (a camera obscura), as well as writing about and researching the motion of planets. According to the annalist Ayni, the Cairo public library housed over two million books during this time period, with six thousand dedicated solely to mathematics and astronomy.[45]

Access to, and the use of, these literary treasures were free to all, and writing materials were afforded gratis. The Caliphs frequently held learned disputations, at which the professors at these academies appeared, divided according to the different faculties: logicians, mathematicians, jurists, and physicians, dressed in their *Khala* or doctoral mantles. The gowns of the English universities still retain the original form of the Arabic *Khala*. The annual revenue of the institutes for the salaries of professors and officials, the provision of teaching materials, and other objects of public scientific instruction was 257 thousand ducats, raised through carefully regulated taxation. In these institutes, they taught every branch of human knowledge.[46]

Cairo under the Fatimids was one of the great medical centres and eventually had three big hospitals: the Ibn-Tulun, one of the first Islamic hospitals, dating to 872; the famous Al-Mansuri, where Ibn al-Nafis was 'Chief of Physicians' in the thirteenth century; and the Qalawun hospital, which was set up by Sultan Qalawun in 1284 and lasted some 650 years until it was demolished in the early twentieth century. Hospitals were models of healthcare for their age, and seemed surprisingly modern in their arrangements. From the records, we know that the hospital was of cruciform shape and divided into wards in which patients were separated depending on the type of illness they were suffering from. The mentally ill were kept apart from those with physical symptoms, and men were housed separately from women. There were also separate units for patients with eye disorders, stomach complaints, and those needing surgery. Hospital doctors by that stage had begun to specialise, and the Qalawun hospital's records tell us that it employed physicians, surgeons, and opthalmologists, as well as administrators, nurses, accountants, and orderlies. According to one report, many new patients were treated everyday. The endowment would also pay for the burial expenses of someone who died during their stay in the hospital.[47]

Like the Abbasids, the Fatimids were patrons of science, medicine, engineering, and learning. They were rationalists to a degree, and could be as hard-core as, say, Al-Mamun. The origins of the Fatimid Caliphs lay in missionary work, and for them, learning was seen as an important vehicle for the spread of faith—not unlike the view taken by Al-Mamun's opponents, the traditionalist scholars of Bagdad. They were keener on astronomy, however, and better builders of institutions. Their most famous legacy, Al-Azhar University, built in 988 initially to train Fatimid missionaries, is one of the world's oldest fully functioning institutions for teaching and research. They were initially mosques designed for teaching, but the curriculum was gradually extended to include a wide range of subjects. Students would travel long distances to study at Al-Azhar.

The Fatimids ruled from 909 to 1171, with Cairo as their capital. In 1171, the famous Salah al-Din Ayyubi (Saladin) overthrew the Fatimids and reunited the forces of Islam sufficiently to drive the Crusaders out. Still, Cairo and Egypt remained relatively safe compared to what was happening in the east of the Islamic world.

Buyid Dynasty (934-1062)

The Buyid dynasty was a Shia Iranian dynasty which mainly ruled over Iraq and central and southern Iran. The Buyid dynasty was founded by Ali Ibn Buya (934-949) popularly known as Imad Al-Doula, who in 934 conquered Fars and made Shiraz his capital. His two brothers, Ahmad Ibn Buya popularly known as Mu'izz Al-Dowla (945-967) and Hasan Ibn Buya popularly known as Rukn Al-Doula (935-976), were the rulers of other territories. Hasan Ibn Buya conquered parts of Jibal and, in 943, managed to capture Ray, which he made his capital. In 945, Ahmad Ibn Buya conquered several territories and officially recognised as the Amir of Iraq. He mainly worked from the central capital, Bagdad (City of Peace).[48] They were all the Amirs under the Abbasid Caliphs. At its greatest extent, the Buyid dynasty encompassed the territory of most of today's Iran, Iraq, Kuwait, and Syria, along with parts of Oman, the UAE, Turkey, Afghanistan, and Pakistan. The Buyids extended a lavish patronage to scientists and scholars.

Azud ud-Dowla (949-983)

Malik Fanakhusrti, popularly known as Azud ud-Doula, is widely regarded as the greatest Amir of the Buyid dynasty, and by the end of his reign, was the most powerful ruler in the Middle East. He was deservedly called the second Augustus of the Arabs. To Azud ud-Dowla, Bagdad owed several hospitals for the sick and refuges for orphans. He built magnificent mausoleums over the tombs of Ali and Husain at Najaf and Karbala.[49]

Azud ud-Dowla, himself a scholar and a mathematician, welcomed to his palace as honoured guests the learned men who flocked to Bagdad from every part of the globe and took part in their scientific controversies. Azud ud-Dowla was an intimate friend of Abdur Rahman Sufi, one of the most brilliant physicists of his age. The great astronomer Abdur Rahman Sufi worked under his patronage and improved the photometry of the stars. Azud ud-Dowla built an observatory in Isfahan where Abdur Rahman Sufi worked. Jaafar, the son of the Caliph Muktafi Billah, made important observations regarding the erratic movements of comets, and wrote a treatise on them; and other princes cultivated the sciences side by side with their subjects.[50]

Azud ud-Dowla is credited with sponsoring and patronising many scientific projects during his time. He also ordered the construction of a great dam between Shiraz and Istakhr in 960. The dam irrigated some 300 villages in Fars province and became known as Band-e-Amir. Among his other major accomplishments was the construction of the Haffar channel, which connected the Karun and Shatt Al-Arab rivers (the Tigris and Euphrates confluence). The port of Khorramshahr was built on the Haffar, at its junction with the Shatt Al-Arab.

A number of astronomers, physicists, and mathematicians flourished under the Buyids, including Al-Kohi and Abul Wafa. Al-Kohi studied and wrote about the movements of the planets. His discoveries concerning the summer solstice and the autumnal equinox added materially to the store of human knowledge. Abul Wafa applied himself chiefly to mathematics and astronomy. His Zij-ush-Shamil (the Consolidated or General Table) is a monument of industry and keen and accurate observation. He introduced the use of the secant and the tangent in trigonometry and astronomical observations. He verified the ancient observations, and discovered, independently of the equation of the centre and the eviction, a third inequality, which is none other than the variation determined six centuries later by Tycho Brahe.[51]

Ghaznavid Empire (977-1186)

Alp Tigin (962-63), a Turkic slave commander under the Samanid monarch Mansur I, began to rule over Ghazna in Afghanistan as a semi-independent governor. However, the founder of the Ghaznavid dynasty was his slave and son-in-law, Subuk Tigin (977-997). The most powerful and important ruler of this dynasty was his son, Mahmud (998-1030), who became independent of the Samanid Empire.

Mahmud of Ghazni (998-1030)

Mahmud was the most successful ruler, greatest conqueror, and unbeatable military general. He extended his empire to the Transoxiana, Afghanistan, Persia, and north-west regions of the Indian subcontinent under the sovereignty of Ghazni. He collected around him a body of scholars and litterateurs who shed a glorious lustre on his brilliant reign. He had the genius of men of merit like Al-Raihan Mohammed Ibn Ahmed Al-Beruni, philosopher, mathematician, and geographer; Firdousi, the prince of poets; Dakiki, and Unsuri.

Al-Beruni's mind was encyclopaedic and his work on astronomy, entitled after his patron Sultan Masud, Al-Kanun-Al-Masudi (Canon Masudicus), is a monument of learning and research. He travelled into India and studied the language of the Hindus, their sciences, philosophy and literature, and embodied his observations in his book *Tahkik-e-Hind.* Al-Beruni communicated to the Hindus the knowledge of the Bagdadian school in return for their notions and traditions. He found among them the remains of Greek science, which had been transported to India in the early centuries of the Christian era, or perhaps earlier, during the existence of the Graeco-Bactrian dynasties. He also wrote on mathematics, chronology, mathematical geography, physics, and chemistry.[52]

Seljuk Empire (977-1186)

The Seljuks were a branch of the Oghuz Turks. Under their leader Tughril, they encountered with the Ghaznavid Empire and defeated them in many fronts. Abu Talib Mohammad Tughril (1037-1063) founded the Seljuk Empire in Iran in his ancestor Seljuk's name. Soon the Arabia, Persia, Rum (Turkey in Asia) and Russia, came under their rule. Under the Seljuk rulers, learning and the arts flourished abundantly. Their grand munificence towards scholarship and science rivalled that of the golden days of the Abbasid rule.

Tughril, Alp Arslan, Malik Shah, and Sanjar, the Seljuk rulers, were notable not only for their greatness of power and clear understanding of what constituted the welfare of their subjects, but also for their intellectual gifts and ardent enthusiasm for the cause of learning. The Seljuks did render invaluable service to culture through their patronage of scientists, historians, poets, physicians, and astronomers. They continued to be the ruling people in its older portions after the downfall of the Seljuk Empire. After the sack of Bagdad by Hulegu, the Seljuks reinstated the Caliphate and became temporal head under the Caliphs.

Jalal ud-Din Malik Shah (1072-1092)

Seljuk ruler Jalal ud-Din Malik Shah (1072-1092) and his Vazir Khwaja Hasan Nizam ul-Mulk collected around them a galaxy of astronomers, poets, scholars, and historians in his capital, Isfahan. The first large-scale observatory in the Muslim .world was that built by Sultan Malik Shah in Isfahan in the late eleventh century. The astronomical observations conducted in his reign by a body of savants, with Umar Khayyam and Abdur Rahman Al-Hazini at their head, led to the reform of the calendar, which preceded the Gregorian by six hundred years and is said by a competent authority to be even more exact. The era which was introduced based on these observations was named after Malik Shah, the Jalalian.[53]

In the interval which elapsed between the rise of Mahmud of Ghazna and the fall of Bagdad, there flourished a number of philosophers and scientists, among whom shone the great Avicenna (Abd Ali Husain Ibu-Sina), Fath Ibn Nabeghah Khakani, Mubashshar Ibn Ahmad, and his son Mohammad.[54]

Vazir Khwaja Hasan Nizam ul-Mulk (1072-1092)

Abu Ali Hasan Ibn Ali Tusi, better known by his honorific title of Nizam ul-Mulk, was a Persian scholar, political philosopher, and Vazir of the Seljuk Empire. He was the *de facto* ruler of the empire for 20 years after the assassination of Sultan Alp Arslan in 1072. He was one of the wisest prime ministers of the century and was a capable administrator and a sagacious statesman. One of his most important legacies was founding madrasahs in cities throughout the Seljuk Empire. He wrote *Siyasat Nama* (Book of Government), a political treatise that uses historical examples to discuss justice, effective rule, and the role of government in Islamic society. His fanatical devotion to the Hanafi doctrines made him harsh towards the heretics, especially the Shi'ites and Ismailies.[55]

He was, however, a liberal patron of men of letters and a sincere friend of scholars. The Nizamia College, which he founded, produced some of the greatest scholars and theologians of the age. Among its professors was the great Al-Ghazali. Nizamul-Mulk was later dismissed owing to the jealousy of the favourite wife of Malik Shah, Turkan Khatun. He did not long survive his disgrace and was assassinated by a young man, a follower of the Assassins (Ismaili Shia), who approached him in the guise of a suppliant and inflicted upon him a mortal wound.[56]

Ahmad Sanjar (1097-1157)

Ahmad Sanjar was the Seljuk ruler of Khorasan from 1097-1118 and became Sultan of the Saljuk Empire from 1118-1157. He was the son of Malik Shah I and took part in the succession war against his three brothers and a nephew and was given the province of Khorasan to rule under his brother Muhammad I. In 1118, he became Sultan by defeating Mahmud II. The Mediaeval sources described Sanjar as having "the majesty of the Khosrows and the glory of the Kayanids". Persian poetry flourished under Sanjar, and his court included some of the greatest Persian poets, such as Muizzi, Nizami Aruzi, and Anvari. The death of Sanjar meant the end of the Seljuk dynasty as an empire, since they only controlled Iraq and Azerbaijan afterwards. Sanjar is considered one of the most prominent Seljuk sultans and was the longest-reigning Muslim rulers until the arrival of the Mongols.[57]

Mongol Empire (1206-1368)

The Mongol Empire emerged in 1206 from the unification of several nomadic tribes of Mongolia under the leadership of Genghis Khan (c. 1162-1227). Hulegu Khan (c. 1217-1265), the grandson of Genghis Khan, attacked the Abbasid Empire and destroyed the city of Bagdad, as well as putting an end to the Abbasid Caliphate in 1258. They swept like overwhelming torrents over western Asia. Wherever they went, they left misery and desolation. Their barbarous campaigns and their savage slaughters put an end for a time to the intellectual development of Asia.

Genghis Khan's grandson Berke Khan (1257-1266) converted to Islam due to the efforts of Saif ud-Din Dervish from Khorazm, thus Berke became one of the first Mongol rulers to convert into Islam. He was a Mongolian military commander and ruler of the Golden Horde (division of the Mongol Empire, who effectively consolidated the power of the Blue Horde and White Horde). The moment the Mongols adopted the religion of the Prophet of Arabia, a change came over them. From the destroyers of the seats of learning and arts they became the founders of academies, the protectors of the learned and built Madrasahs and observatories.[58]

Almost all Mongol Amirs and rulers protected philosophers and scientists and patronised the learning of science and philosophy. Oljaitu Khan, known as Sultan Khoda-Bendah (1280-1316), sixth in descent from Genghis, was distinguished for his attainments and his patronage of the sciences. The Mongols protected philosophers like Nasir ud-Din Tusi, Muwayyad ud-Din Al-Orezi of Damascus, Fakhr ud-Din Al-Maraghi, Mohi ud-Din Al-Maghribi, Ali Shah Al-Bokhari, and many others. The successors of Hulegu tried thus to restore to Islam what their ancestor had destroyed.[59]

Timurid Dynasty (1370-1507)

The Timurid dynasty was founded by Timur Gurkani, or Tamerlane of Samarqand. It was a dynasty or clan of Turko-Mongol origin. Its origin goes back to the Mongol tribe known as the Barlas in Uzbekistan, who were remnants of the original Mongol army of Genghis Khan. The word Gurkani derives from the Mongolian word Kuragan, meaning son-in-law. This was an honorific title used by the dynasty as the Timurids were in-laws of the line of Genghis Khan as Timur had married Saray Mulk Khanam, a direct descendant of Genghis Khan.

Timur Gurkani (1370-1405)

Amir Timur, or Timur Gurkani, was a Turko-Mongol conqueror who founded the Timurid Empire in and around Afghanistan, Iran, and Central Asia. As an undefeated commander, he is widely regarded as one of the greatest military leaders and tacticians in history. He was a patron of science and poetry, himself fond of the society of the scholars and artists of his day, an author, as well as a legislator of no mean order. Magnificent colleges, splendid mosques, and vast libraries testified to the taste for letters of this remarkable man. The college founded by his consort, Bibi Khanam, and known by her name, still strikes the observer as one of the most imposing and most beautiful products of Saracenic architecture.[60]

Ulugh Beg (1447-1449)

Timur's grandson Ulugh Beg, the ruler of Uzbekistan, maintained the literary and scientific glories of Samarkand. He built his capital, Samarkand one of the most important scientific centres of the fifteenth century, a time once considered to be barren. He was the patron of the Ghiyath al-Din Jamshid Masud al-Kashi, a great astronomer of that time, who took part in the construction of the great observatory. Himself an astronomer of a high rank, he presided over the observations which have immortalised his name.

Astronomical research carried out at Ulugh Beg's observatory produced a star catalogue consisting of 1,018 stars. Astronomers and mathematicians of the fifteenth century made significant discoveries in the Madrasah of Ulugh Beg in Samarqand. Ulugh Beg is separated by only a century and a half from Kepler, the founder of modern astronomy.[61]

Islam inaugurated the reign of intellectual liberty. In fact, observatories, hospitals, madrasahs, and public libraries were four characteristic institutions of Islamic civilization. It has been said that as long as Islam maintained its pristine character, it was a warm protector and promoter of knowledge and civilization, as well as a zealous ally of intellectual freedom. The moment extraneous elements attached themselves to it; it lagged behind in the race of progress.

Mamluk Sultanate (1250-1517)

The Mamluk Sultanate, also known as the Mamluk Egypt or the Mamluk Empire, was a state that ruled Egypt, the eastern Mediterranean region of western Asia and the Hejaj, or western Arabia. It was ruled by a military caste of Mamluks (manumitted slave soldiers), at the head of which was the sultan. The Abbasid Caliphs were their nominal sovereigns. The Sultanate was established with the overthrow of the Ayyubid dynasty in Egypt in 1250 and was conquered by the Ottoman Empire in 1517.

The Mamluks were famous for their architecture in Cairo, as the main centre of patronage, and many other cities such as Damascus, Jerusalem, Aleppo, and Medina. Patrons, including sultans and high-ranking amirs, typically set out to build mausoleums for themselves but attached to them various charitable structures such as madrasahs, khanqahs, sabils, mosques etc. The decoration of monuments also became more elaborate over time, with stone-carving and coloured marble panelling and mosaics (including *Ablaq*) replacing stucco as the most dominant architectural decoration. Monumental decorated entrance portals became common compared to earlier periods.

Ottoman Empire (1517-1922)

The Osmanli Turks, or Ottoman Turks, were a small clan of the Oghuz tribe founded by Osman. They were driven westward by the Mongol migration, which had taken refuge in Asia Minor early in the thirteenth century. The victories of Maritza (1364), Kosovo (1389), and Nicopolis (1594), gave the Turks the whole of the Balkan Peninsula. The Crimea was annexed in 1475 and Salim I, the Grim, took Kurdistan from the Persians and Syria, Egypt, and Arabia from the Mamluks in 1517. Salim I became the master of the holy cities of Makkah and Madinah, and the last Abbasid Caliph of Cairo transferred to him the Caliphate of Islam. From this time onward, the Ottoman Turks assumed the responsibility of the Caliphate.[62]

At the same time, there was a Shia kingdom in Persia under Ismail Shah[63] and both were always at war. The bitter strife between Turkey and Persia destroyed the unity and solidarity of Islam. As for intellectual or artistic progress, the Turkish rule was not far behind in this field. The Ottoman Empire, which in the days of its splendour embraced the whole of Southeast Asia and a considerable portion of Europe and Africa, revived afresh the glory of the old Caliphate. The Ottomans translated many Arabic texts into Turkish and produced new works. They further rearranged old material, corrected geographical information, and added new observations. For example, Abu'l Fida's *Taqwim al-Buldan* (A Sketch of the Countries) was translated into Turkish by Sipahizade Mehemmed bin Ali (d. 1588), who supplemented and rearranged the material in alphabetical order. Turkish geographers also produced considerable new literature on marine geography and navigation.[64]

Seyyidi Ali Re'is bin Husain (d. 1562), also known as Katib-e-Rumi, wrote a book on the Indian Ocean entitled *Al-Muhit*, using the experiences of South Arabian sailors, some of whom had served as guides to Vasco de Gama on his voyage to Calicut. Piri Muhyi'l Din Re'is (d. 1554) produced a world map in 1513, for which he used as sources maps containing Portuguese discoveries up to 1508, and another map containing discoveries by Christopher Columbus during his third voyage (1498). He had obtained this map from a Spanish sailor who had voyaged with Columbus to America three times and who had been made a Turkish prisoner in 1501 at Valencia by none other than Piri Re'is's uncle, the famous naval hero Kemal Re'is. One of the most comprehensive geographical works of the early seventeenth century, written by Mustafa bin Abdallah, popularly known as Haji Khalifa (1609-1657), also used Muslim as well as at least one European source, the Atlas Minor (1621) of Gerhard Mercator.[65]

As for educational development in the Ottoman Empire, the Ottoman medrese (madrasah) was the institution responsible first and foremost for acquiring deeper knowledge in the field of religion. Mosques and libraries also functioned under certain circumstances as possible schools. Over the course of the fourteenth century, the first Ottoman madrasahs were established, and it became possible to obtain Islamic knowledge locally. By the time of Mehmed II (1444-1446 and 1451-1481), the conqueror of Constantinople and founder of Istanbul University, a local tradition of religious studies had already begun to develop. To this, Mehmed II added a hierarchy of the various teaching institutions throughout the Ottoman state. This development continued vigorously, and later on, the madrasahs that Suleyman (1520-1566) included in his complex outranked Mehmed's as the most prestigious. However, as time passed and the need to acquire broader knowledge increased, it offered a broad education as well. For those who desired a wider understanding, some madrasahs could also provide the comprehensive knowledge required by polymaths.

This was the impression of one eighteenth-century observer, G.B. Toderini, a learned Jesuit who lived in Istanbul from October 1781 to May 1786. He described the grandness of Ottoman madrasahs (he referred to them as "accademia"), noting generally how magnificent madrasah buildings were and how impressive the professors were. He then elaborated on some of the more famous and centrally located madrasahs in Istanbul at that time, organised according to the chronology of their establishment, from the schools of Mehmed II to those of Abdul Hamid I (1774-1789). More importantly, Toderini described Ottoman madrasahs as institutions devoted to all the sciences, where religious studies such as *Tafsir* and *Hadith* were taught alongside astronomy and geometry.[66]

Indeed, solid evidence exists that astronomy and mathematics were sometimes taught in some Ottoman madrasahs, and even in the Sufi lodges of *Tekkes* and *Zaviyes*. Geometry and astronomy were mentioned as separate courses in the curriculum; algebra was usually part of arithmetic rather than an independent field of study. Some historians, mainly Turkish ones, suggest that non-religious studies (including arithmetic, geometry, astronomy, music, chemistry, medicine, agriculture, and geography) were almost (but not quite) as important as religious studies at the Madrasah. However, the extent to which non-religious subjects were routinely taught in the Madrasahs and their relative importance in the curriculum can still be questioned.[67]

The activities of the medical school and hospital in the Suleymaniye complex in Istanbul were extraordinary rather than the rule. There may have been a few other medical schools; for example, the Madrasah at the Edirne complex of Bayezid II (1481-1512) may have actually been a medical school. The visitors highly praised the level of the teaching, as well as the students and the extent of their dedication to healing people. However, only a handful of madrasahs in the Ottoman Empire specialised in medicine or taught medicine in addition to bodies of religious knowledge. There were hardly any professional medical schools, just as there were no vocational schools for many other professions, prior to the educational reforms of the nineteenth century.[68]

In the eighteenth century, in 1773, the school of mathematics was opened for the navy. It served as a model for similar engineering schools for the army in 1793, medical academies in 1827, 62 music schools in 1831, war sciences studies in 1834, and other fields. Mainly, the military schools gave the impetus for a new, broad education system. The nineteenth-century education reforms created a learning environment that was new in its content, its socio-cultural atmosphere, and its physicality. New schools were built throughout the empire. They were no longer necessarily located adjacent to mosques, and their physical organisation was different from the traditional setting of the mekteb.[69]

Schools were now structured according to the hierarchical model of the West. The new educational model was a state-governed system: educational and pedagogical techniques were decided upon at the centre and implemented through the provinces. During the last decades of the Ottoman Empire or at the beginning of the twentieth century, the expansion of education included female students as well. Traditional schools, the mektebs (Maktabs, primary schools) and madrasahs, continued to exist side by side with the new system, and continued to attract many in Ottoman society.

The Ottoman Empire was essentially a military despotism, and the Turks did not largely distinguish themselves as patrons of the arts and sciences. For education, they were confined to madrasas and maktabs. They gave complete religious toleration to Christians and Jews, and allowed them self-government in the management of their communal affairs. But the intellectual development was lacking, as the Turks did not devote sufficient attention to the economic development and intellectual growth of their empire.

NOTES

1. *Sunan al-Tirmidhi*, Number 3723.
2. *Ibid.* Number 2687.
3. Nasr, *op. cit.,* p. 12.
4. *Ibid.*, p. 16.
5. *Ibid.*
6. *Ibid.*, pp. 22-23.
7. *Ibid.*
8. Bhargava, *op. cit.,* pp. 36-38.
9. Masood, Ehsan (2009). *Science & Islam: A History of Icon Books*, London, pp. 3-4. The city of Bagdad was constructed in the middle of the 8th century at the command of the Abbasid Caliph, whose faction had long been in conflict with a rival Islamic faction, the Umayyads. Bagdad would become a jewel in the crown of the Islamic world, with a fully functioning political bureaucracy and a house of wisdom, established in the year 830. Dorsey

Armstrong (2014). *Great Minds of the Mediaeval World*, Virginia, p. 58. Muslim writers, such as Athir, Fakhari, and Masudi, inform us that the people believed that the Abbasid Caliphs would last for ever, and an idea was prevalent in men's minds that if an Abbasid Caliph were killed, the order of the Universe would be disturbed, the sun would be darkened, the rains would stop, and vegetation would dry up. Bhargava, *op. cit.,* p. 49.

10. Bhargava, *op. cit.,* p. 94.
11. Hoodbhoy, Pervez, "*How Islam Lost Its Way: Yesterday's Achievements Were Golden: Today, Reason Has Been Eclipsed*," Washington Post, December 30, 2001.
12. Iqbal, Muzaffar, *op. cit.*, p. 153. The beauty and splendour of the city, before its sack by the Mongols, have been immortalised in glowing lines by Anwari, the most brilliant of panegyrists:
    "Blessed be the site of Bagdad, seat of learning and art,
    None can point in the world to a city equal to her,
    Her suburbs vie in beauty with the blue vault of heaven,
    Her climate in quality equals the life-giving breezes of heaven,
    Her stones in their brightness rival gems and rubies,
    Her soil in beneficence has the fragrance of the amber,
    The morning breeze has imparted to the earth the freshness of Taba (the tree of Paradise),
    And the winds have concealed in her water the sweetness of Kausar (the spring of Eden),
    The banks of the Tigris with their beautiful damsels surpass Khullakh (A city in Cathay famous for the beauty of its women),
    The gardens filled with lovely nymphs equal Cashmere,
    And thousands of gondolas on the water,
    Dance and sparkle like sunbeams in the sky."
    Ameer Ali, *op. cit.*, pp. 368-69.
13. Ameer Ali, *op. cit.*, p. 369.
14. *Ibid.*, p. 370.
15. *Ibid.* An Indian astronomer who arrived in the court of the Abbasid Caliph Al-Mansur as part of a delegation from Sind, for instance, was probably a Hindu. He knew Sanskrit and helped Al-Fazari translate a Sanskrit astronomical text into Arabic; this text contained elements from even older astronomical traditions. Iqbal, Muzaffar, *op. cit.*, p. 13.
16. *Ibid.*, p. 371.
17. *Ibid.*, p. 374.
18. Dahlan, Winai, *op. cit.*, p. 89.
19. Ameer Ali, *op. cit.*, p. 374.
20. *Ibid.*, pp. 375.
21. Al-Masudi (896-956) saw this world map and commented, "It depicted the universe with spheres, the stars, land and the seas, inhabited and barren (regions of the world), settlements of peoples, cities." Cartography after the tenth century developed marked Islamic features, showing the influence

of the Islamic worldview in various realms such as politics and culture, as well as various spiritual aspects of Islam. It emerged out of a new tradition in Islamic geography. Iqbal, Muzaffar, *op. cit.*, pp. 46-47.

22. Ameer Ali, *op. cit.*, pp. 373 & 375.
23. *Ibid.*, p. 371.
24. *Ibid.*, p. 375.
25. Iqbal, Muzaffar, *op. cit.*, p. 153. (Ameer Ali quotes Gibbon, "For three days the streets ran with blood, and the water of the Tigris was dyed red for miles along its course. The horrors of rapine, slaughter, and outraged humanity lasted for six weeks. The palaces, mosques, and mausoleums, were destroyed by fire or levelled to the earth for their golden domes. The patients in the hospitals and the students and professors in the colleges were put to the sword. In the mausoleums the mortal remains of the Sheikhs and pious Imams, and in the academies, the immortal works of great and learned men, were consumed to ashes; books were thrown into the fire, or, where that was distant and the Tigris near, were buried in the waters of the latter. The accumulated treasures of five centuries were thus lost for ever to humanity. The flower of the nation was completely destroyed.") Ameer Ali, *op. cit.*, p. 402.
26. Bhargava, *op. cit.*, p. 51.
27. The Mongols were originally pagans, and Kublai Khan was the first Mongol of the royal house to become Buddhist, while Berke Khan (1257-1266), Taqudar Khan (1282-1284), and Ghazan Khan (1295-1304) were the first to embrace Islam. When the Mongols burst into Muslim countries, Islam seemed lost. It was the most critical period in the history of Islam, as it was fiercely attacked on one side by the army of the fanatical crusaders, who were fired with religious ardour, and on the other by the wild plundering hordes of the Mongols. It seemed as if the barbarians had become the masters of Asia, but Islam survived these disasters. Islam won a dazzling victory when Ghazan Khan, the seventh ruler, accepted Islam and entered into friendly relations with Muslim princes. Bhargava, *op. cit.*, p. 139.
28. The name Tatar first appeared among nomadic tribes living in north-eastern Mongolia and the area around Lake Baikal in the fifth century. Unlike the Mongols, these peoples spoke a Turkic language, and they may have been related to the Cuman (nomadic Turkish people) or Kipchak (a loosely organised Turkic tribal confederation) peoples. After various groups of these Turkic nomads became part of the armies of the Mongol conqueror Genghis Khan in the early thirteenth century, a fusion of Mongol and Turkic elements took place, and the Mongol invaders became known to the world as the Tatars. After Genghis Khan's empire broke up, the Tatars became especially identified with the western part of the Mongol domain, which included most of European Russia and was called the Golden Horde. These Tatars were converted to Sunnite Islam in the fourteenth century. These Tatars consisted of most of the army of Tamerlane. Owing to internal divisions and various foreign pressures, the Golden Horde disintegrated

late in the fourteenth century into the independent Tatar Khanates of Kazan and Astrakhan on the Volga River, Sibir in western Siberia, and Crimea. *https://www.britannica.com/topic/Tatar.*

29. Masood, Ehsan, *op. cit.*, p. 7.
30. *Ibid.*
31. Bhargava, *op. cit.*, pp. 163-64.
32. *Ibid.*, pp. 146-48.
33. *Ibid.*, pp. 148-49 & 155.
34. Ameer Ali, *op. cit.*, pp. 378-79. Of Cordova, Ameer Ali quotes an English writer who says, "Beautiful as were the palaces and gardens of Cordova, her claims to admiration in higher matters were no less strong. The mind was as lovely as the body. Her professors and teachers made her the centre of European culture; students would come from all parts of Europe to study under her famous doctors, and even the nun Hroswitha far away in her Saxon Convent of Gaudersheim, when she told of the martyrdom of Eulogius, could not refrain from singing the praises of Cordova, the brightest splendour of the world. Every branch of science was seriously studied there, and medicine received more and greater additions by the discoveries of the doctors and surgeons of Andalusia than it had gained during all the centuries that had elapsed since the days of Galen (Jalinus). . . . Astronomy, geography, chemistry, natural history, all were studied with ardour at Cordova; and as for the graces of literature, there never was a time in Europe when poetry became so much the speech of everybody—when people of all ranks composed those Arabic verses which perhaps suggested models for the ballads and canzonettes of the Spanish minstrels and the troubadours of Provence and Italy. No speech or address was complete without some scrap of verse, improvised on the spur of the moment by the speaker or quoted by memory from some famous poet."
35. Bhargava, *op. cit.*, p. 159.
36. *Ibid.*, pp. 165-67.
37. *Ibid.*, p. 169. (Bhargava quotes from Dozy's Spanish Islam, "Never had so learned a prince reigned in Spain, and although all his predecessors had been men of culture who loved to enrich their libraries, none of them had sought so eagerly for rare and precious books. At Cairo, Bagdad, Damascus, and Alexandria, Hakam had agents who copied or bought for him—grudging no cost—ancient and modern manuscripts. With these treasures his palace overflowed on all sides, too, were to be seen copyists, binders, and illuminators." Dosy further says, "The catalogue of the Caliph's library occupied forty four volumes, of which each comprised twenty sheets—or fifty according to some chroniclers—though they contained only the titles of the works. Some writers assert that the number of volumes amounted to four hundred thousand. All of these volumes Hakam had read, and most of them he had annotated. At the beginning or end of each book he wrote the name, surname, and patronymic of its author, his family and tribe, with the year of his birth and that of his death, and anecdotes relating to

him. These notices were precious; no one was more learned in literary history than Hakam, and his annotations were always held authoritative by Andalusian scholars." Bhargava again quotes Dozy, "Hakam's liberality to the learned, whether Spaniards or foreigners, was indeed boundless, and they accordingly flocked to his court. The monarch encouraged and protected them all—even philosophers, who were at last enabled to devote themselves to their studies without the fear of being massacred by bigots.") Bhargava, *op. cit.*, pp. 168-169.

38. *Ibid.*, pp. 179-80.
39. *Ibid.*, p. 171.
40. *Ibid.*, p. 137.
41. *Ibid.*
42. Abu Tahir Carmaathi, with an army of six hundred horsemen and nine hundred infantry, entered the sacred city of Makkah, killed, plundered, and imprisoned the Makkans, and carried off the Black Stone and other sacred relics. In this appalling atrocity, thirty thousand Muslims were slain, of whom one thousand nine hundred met their death in the precincts of the Kaabah itself. Bhargava, *op. cit.*, pp. 111-12 &122.
43. Bhargava, K.D. (1961). *A Survey of Islamic Culture and Institutions*, Allahabad. pp. 324, 337 & 372; see also Masood, Ehsan, *op. cit.*, pp. 10-11.
44. *Ibid.*, p. 377.
45. *Ibid.*, pp. 377-78.
46. *Ibid.*, p. 337.
47. Masood, Ehsan, *op. cit.*, pp. 11-12.
48. *https//en.m.wikipedia.org/wiki/Buyid_dynasty.*
49. Bhargava, *op. cit.*, p. 376.
50. *Ibid.*
51. *Ibid.*
52. *Ibid.*, p. 380.
53. *Ibid.*, p. 381.
54. *Ibid.*, p. 382.
55. *Ibid.*, p. 74.
56. *Ibid.*
57. *https://iranicaonline.org/articles/sanjar.*
58. Ameer Ali, *op. cit.*, p. 382.
59. *Ibid.*, p. 383.
60. *Ibid.*
61. *Ibid.* (A certain number of Madrasahs in the Muslim world were special institutions where traditional sciences were accompanied by a thorough study of various disciplines of science. These institutions served the dual purpose of being the centres of religious as well as natural sciences. The still-standing, fabulous madrasah of Ulugh Beg in Samarqand is one such example. The high levels of religious scholarship accompanied by an equally high level of scientific investigations were taking place at this madarsah. Ghiyath al-Din al-Kashl, the most accomplished mathematician

and astronomer of that time, wrote a letter to his father shortly after his arrival in Samarqand around 1420. The letter speaks of the "Sultan (that is, Ulugh Beg) being extremely well-educated in the (sciences of) the Quran, in Arabic grammar, in logic, and in mathematical sciences". It mentions the presence of sixty or seventy astronomers and mathematicians in Samarqand at the Sultan's madrasah. Another account states that "at the time Ulugh Beg's observatory flourished, it was carrying out the most advanced observations and analysis being done anywhere. In the 1420s and 1430s, Samarqand was the astronomical and mathematical capital of the world".) Iqbal, Muzaffar, *op. cit.*, p. 231.

62. Bhargava, *op. cit.*, p. 141.
63. Shah Ismail, the ruler of Persia, made Shi'ism as state religion and persecuted the Sunnis of his state. He ordained the public cursing of the first three Caliphs of Sunnis, Abu Bakr, Umar, and Osman, and decreed that all who heard the cursing should respond, "May it be more, not less." Those who refused to give this response were to be killed. He also instituted the public cursing of Abu Bakr, Umar and Othman in the streets and markets and threatened recalcitrants with punishment. Shia theology was encouraged and the doctrines of the Shia creed were expounded by able, earnest, and pious theologians of Persia. This practice continued till the end of the Kajar dynasty and the beginning of the Safawi dynasty. Persian genius blossomed forth under the Safawi dynasty, and the country's progress was exemplified by the extension of trade and commerce, the patronage of scholars and holy men, and the expansion of the empire. It is unfortunate for Islam that the two great Muslim countries, Persia and Turkey, have carried on a fierce struggle for centuries and have never been able to combine against a common foe. Bhargava, *op. cit.*, pp. 128-29.
64. Iqbal, Muzaffar, *op. cit.*, p. 46.
65. *Ibid.*
66. Miri Shefer-Mossensohn (2015). *Science among the Ottomans: The Cultural Creation and Exchange of Knowledge*, Austin, p. 61.
67. *Ibid.*
68. *Ibid.*, p. 67.
69. *Ibid.*, p. 81.

5
Science and Rationalism in Islamic Theory and Practice

Islam as a religion not only guides people to believe in Almighty Allah, His Realm, and His Prophet, but it also encourages people to study His creation, or the Heaven and Earth, which eventually leads to science, natural history, philosophy, reason, and rationalism. Rationality, or rationalism, is the exercise of reason in thought or action, which is very important for investigation and research. Logic is also associated with reason or rationalism. Rationalism needs well-grounded premises: that the factual bases and principles the rational person uses are known to be correct or can be accepted as correct for good reasons; sound logic; that these principles are used in accordance with the laws of logic; and systematic and controlling resort to reason; that such use of well-grounded principles and sound logic is the basic method by which the person determines his beliefs, makes his arguments, or decides on his actions.

From the seventh to the seventeenth century, Islam continuously expanded in its following as well as physical boundaries. The rulers and intellectuals of the Islamic empires promoted science and rationalism by following the Quranic and Prophetic guidance. We could see that scientists who made the Islamic scientific tradition possible during the eighth and sixteenth centuries were more frequently also the authors of religious texts of an advanced nature. Muslim scientists of the pre-seventeenth century era did not see their religion and their science as two separate entities. Science and scientists help religious institutions and authorities to prepare for these rituals and celebrations by furnishing the required astronomical information. In Islam, not only is the annual cycle of rituals and celebrations dependent on astronomically determined times but also daily practices, such as the five obligatory prayers[1] and fasting[2].

Islam and Its Expansion

Islam is a religion that was spread and promoted by the Prophet Mohammad (pbuh), revealed by Allah through the Quran, for human being. The Unity of Allah, universality, equality, and brotherhood extended to all Muslims, are the best features of Islam. In Islam, unlike other religions, there is no concept of intermediary or the priestly class between man and Allah. The Quran and *Hadith* are enough to understand the teachings of Islam. Islam gives the concept of equality both in theory and practice so far as Muslims are concerned and there is no distinction between man and man on the plane of humanity, not only in the formal act of worship but also in daily social intercourse.[3]

Islam also opposes economic inequality and considers it the most dangerous for social solidarity and a cause of discontent, unrest, division of society and castes. Islam always promotes seeking knowledge of heaven and earth and Almighty Allah, the Creator of the heaven and earth has made man as its trustee. Accordingly, the theologians have given emphasis to research, investigation, and science.

Military Expansion

When the Prophet Mohammad (pbuh) was born, the Sassanid Empire of Persia and the Byzantine Empire of Constantinople were two great Empires which were always at war. Byzantine Empire had reached its greatest extent and expanded into the northern Arabian Peninsula, including Syria, Jordan, Lebanon, Palestine, Iraq, Anatolia, Egypt, North Africa, Morocco, and Spain. However, the geography and boundaries of these empires were changed within a century of the Prophet's death. Islam had spread throughout the Middle East, much of Africa and Asia, and as far as Narbonne in southern France.

This expansion of the geographical boundaries of the Muslim world occurred in two successive waves. The first took place between 632 and 649 and the second between 693 and 720. As history shows, Islam as a religion spread more quickly than other religions, and as an Islamic state or empire, it is growing at a similar rate. It was the inherent truth of the message of Islam, rather than military victories, which established Islam across this vast region. It was really miraculous that a few Arab armies, commanding at most a few thousand soldiers, passing through certain selected routes of the ancient world, could bring such a large number of countries and their people into the fold of Islam.

In 633, the Battle of the Chains, followed by other victories, admitted the Muslims to Chaldaca (Irak-Arabi) and gave them the city of Hira. In

634, the battle of Yarmuk opened Syria to their arms; Damascus fell in 635, Emesa, Antioch, and Jerusalem in 636; and the conquest of Caesarea completed the subjugation of Syria in 638. Meanwhile, the victory of Kadisiya in 635 was followed by the conquest of Madain (Seleucia-Ctesiphon); the old capital of Chaldaca was acquired in 637; Mesopotamia was subdued, and the cities of Basra and Kufa were founded; and Khuzistan and Tustar were annexed in 638-40. The decisive battle of Nahawand in 642 put an end to the Sasanid dynasty, and gave the whole of Persia to Muslims. By 661, the Muslims were at Herat and soon carried their arms throughout Afghanistan and as far as the Indus, where they established a government in Sind. In 674, they occupied Bokhara, and two years later, Samarkand, but these early raids in Transoxiana were not converted into settled conquests until 711. In the East, the Caliphate reached its utmost limits in little more than forty years after the Muslims had first led a campaign outside Arabia. The phenomenal progress of Muslims has never been equalled in any period of world history.[4]

In the West, the progress of Muslim arms was slower. In 641, Egypt was conquered, and by 647, the Barbary Coast was overrun up to the gates of Roman Carthage; but the wild Berber population was more difficult to subdue than the pleasure-loving subjects of the Sasanids of Persia or the sophisticated and logic-chopping Greeks of Syria and Egypt. The Kayrawan was founded as the African capital in 670; Carthage fell in 693, and the Arabs pushed their arms as far as the Atlantic with ruthless energy. From Tanjier, they crossed into Spain in 710, and the conquest of the Gothic kingdom was completed with the fall of Toledo in 712. In 726, southern France was overrun, and in spite of the victory of Charles over the Arabs near Tours in 732, Muslims continued to hold Narbonne and to ravage Burgundy and the Dauphine for a long time. Thus, less than a century after the conquest of Iraqi Arabia, the Caliphate reached its greatest extent in the West.[5]

To the north, the Greeks retained Anatolia, which never belonged to the Caliphate, but the Muslims invaded Armenia and reached Erzurum in about 700; Cyprus had been annexed as early as 649, and Constantinople was several times besieged from 670 onwards. Thus the empire of the Caliphs at its widest extended from the Atlantic to the Indus and from the Caspian to the cataracts of the Nile.[6]

Expansion of Islamic Belief: Islam, thus, is not merely a set of beliefs and associated practices, but a way of being, a path that a traveller takes with a definite destination in mind. Moreover, it is a path that transforms travellers as they order their lives according to Divine guidance. While there is, of course, a set of beliefs associated with Islam, these beliefs are not merely abstract ideas; they have been presented in concrete form through the Sunnah of the Prophet of Islam, which constitutes the real-life model for travellers on the religion of Islam. All these beliefs and ideas developed the thought of rationalism and the study of nature or science.

The religion Islam paved the way of free thinking and study of nature that resulted in Islamic science, natural philosophy, or rationalism. Understanding of the relationship between science and Islam depends on how one defines the central purposes and the appropriate boundaries of science and religion. Islam is not a fluid conceptual framework that keeps changing with time; rather, an Islamic way of being can be verifiably traced back to the life of the Prophet of Islam, a life lived in the full light of Quranic instructions, or rather, a living model of the Quran. Thus, within the broad contours of Islamic civilization, all kinds of rulers, patrons of learning, scholars, and scientists have existed and continue to exist.

The Islamic Empire consisted of a society that was multicultural in terms of languages, customs, traditions, and religion. As Muslims went forth from Arabia to conquer the countries surrounding them, they encompassed vast lands with peoples of different faiths and cultures. Within the first hundred years, a very large number of micro-transformations took place in the ancient lands of Asia, Africa, and Central Asia to give birth to a new social and cultural mosaic. This multi-cultural, multi-ethnic, and even multi-religious society was to create the conditions for Islamic civilization's seven-century-long scientific enterprise. This vast social change in the eighth century was the result of mass conversions, migrations, the establishment of new cities and new administrative and fiscal institutions, and a new social contract that Islam established with Jewish and Christian communities.

Muslim scientists followed the Quranic description of the earth and heaven and studied during the period from the eighth to the sixteenth centuries, believing that the Earth was created by Almighty Allah for a fixed duration and for a definite purpose. For these scientists, it was a place of wonders and of observations that led to an understanding of Divine Wisdom, Power, and Mercy, with sustenance provided for every living creature on Earth. For these scientists, the Quranic descriptions of various processes, such as the regeneration of water and of the Earth's coming back to life after having been barren (Quran, 36:23) were observable realities that directed human reflection to the Creator.

The Holy Quran many times says that the earth and heaven are made for mankind and they should study and utilise the resources, *Hadith* directs the same. These are the branches of science that were directly related to

Islamic practices: astronomy, which was used to determine the distance and direction towards Makkah (the direction Muslims face for their obligatory prayers five times a day); geography; geodesy; cartography; *Mawaqeet* (the science of timekeeping); and other such branches of science that have a direct relationship with Islamic practices. These are not simply cases of "science in the service of religion," as is sometimes claimed; rather, these sciences emerged from a specific view of nature anchored in Islam.[7]

The expansion of geographical borders and the subsequent cultural synthesis forced the keenest minds of the times to continuously formulate answers to a wide range of questions arising from. The rapid geographical expansion also posed new questions related to the practice of faith and that required immediate attention: how to determine the correct *Qiblah* (direction towards the Kaabah in Makkah) for the five obligatory daily prayers from a distant city; how to calculate the correct amount of *Zakah* (obligatory charity on wealth and material goods); how to apply the principles of inheritance outlined in the Quran to complex situations. This and numerous other issues arising out of a rapidly expanding social, political, and economic landscape produced a fervour of intellectual activity that resulted in the emergence of new branches of knowledge.

Thus, the Islamic Empire not only consisted of Muslims from three continents, Arabs, Persians, and Turks, Africans, Indians, and other Asians, but also Jews, Christians, and other faiths. Therefore, scholars from all faiths worked under the umbrella of Islam to produce a unique culture of knowledge and learning. The religious instructions, such as the Quranic conception of human responsibility, developed so strongly. The Quranic passages in which human responsibility and the freedom of human will are laid down show that man, within the limited sphere of his existence, is absolute master of his conduct. He is responsible for his actions and for the use or misuse of the powers with which he has been endowed. Free will and freedom to work established a new modern conception of rationalism in society and developed in the fields of learning, philosophy, and science.

Rationalism under Abbasids

During the Abbasid period, science and research fully flourished. The scholars not only studied the Quranic messages and Traditions of the Prophet but a large amount of scientific and philosophical literature brought to the Islamic tradition by the translation movement. This literature included specific cosmological schemes of the Greeks, the Persians, and the Indians. The most prominent among these sciences was cosmology, which was then viewed, examined, and contrasted with the Quranic cosmology by Muslim thinkers; this produced a large body of literature within the Islamic philosophical, scientific, and mystical traditions.

The process of scientific, philosophical, and religious progress went on for several centuries, producing new and refined schemes. In essence, it was an attempt to find ways to harmonise the received data and cosmological theories with the Islamic concept of the cosmos, although the process was complex and often fraught with tension as it had the influence of Aristotle[8] and his school. It developed a rationalistic approach to thinking and created various schools of thought. The rational ideology was against the traditionalist views.

The philosophers of this period were divided into three schools. The materialists, who rejected the Creator and contended that the world has existed from all eternity. Then were the Deists, the Tabiis or Naturalists, who conceded the need for a Creator, who, in their opinion, was more like a machine; they therefore, rejected a future life. Finally, there were the Theists, who were puritans and believed in the Quran and the Commandments of Allah. During the Abbasid period, the second and third groups were in dominance. They were scientists, men of letters, high-grade philosophers, and theologians. Their knowledge was promoted by the Caliphs and their free will made the science and learning more powerful.

The traditionalists, or theists, relied on faith and shunned reason; the rationalists, or naturalists, glorified reason and had little use for faith. Whereas rationalists criticised traditionalist interpretations for intransigence *vis-à-vis* the use of reason, traditionalists claimed that rationalist interpretations could lead to a loss of belief. Reason or rationality is the systematic and controlling use of beliefs, arguments, or actions based on well-grounded premises and valid arguments such that another person who has access to the same information and can understand the argument correctly ought to agree that the premises are well-grounded, that the logic is sound, and that the resultant beliefs, arguments, or actions are correct.

Religious doubt may have partly stemmed from the fact that rationalists encouraged believers to approach Allah directly without interference from religious elites. Traditionalist religious elites were opposed to such rationalist interpretations as well as to the scientific mindset believed to be at their root. Among the Abbasids, Caliphs Al-Mamun gave much emphasis to the translation of the Greek manuscripts and rationalistic approach of Islamic theology as well as scientific development.[9]

The first hundred years of the Abbasid Caliphate were the golden years of Muslim culture, and the period witnessed phenomenal progress in the empire. Under the Abbasid policy of liberalism and free will, philosophy and science were studied with enthusiasm by all sections of the intelligentsia. Following the teachings of Islam and rational freedom, the urge for knowledge and science fully flourished under the supervision of the Caliphate rulers. The Abbasids supported these efforts as long as they were limited to intellectual and literary activity, scientific research, systematic organization, translations from Syriac and Greek languages, and were leaders of the Renaissance.[10]

The Abbasid rulers and intellectuals were aware that Islam always encourages the search of the universe in the hope of uncovering secrets, signs of the Creator, and physical laws governing the universe. Greek philosophy, Zoroastrianism, Manichaeism, the old heathen beliefs of Harran, Judaism, Christianity, all were studied with great interest at a time when the liberal policy of the earlier Abbasids gave free scope to intellect. In the fifth century, an academy of medicine and philosophy had been founded at Gondeshapur in Khuzistan. One of the directors of this institution was summoned to prescribe for Al-Mansur, and from that time the academy furnished a succession of court physicians to the Abbasids. Muslims regarded the new culture as a solvent through philosophical speculation, and Aristotle and Plato, Euclid and Ptolemy, Galen (Jalinus), and Hippocrates produced a profound influence on Muslim life and thought.[11]

Abbasid Caliph Al-Mansur was the ablest of all the Abbasid Caliphs. He believed firmly in and acted upon the principles of religious toleration, and this had most beneficial effects on the mental life of the empire. Christians, Jews, Zoroastrians, and the Heathens of Harran cooperated actively in the supremely important task of interpreting Greek science and culture to the Arabs. The mental food upon which the Arabs were nourished came from different sections and races of Asia. The heathen Syrian source, which had its centre at Harran, played an important part in the propagation of Greek culture, though the information at our disposal is limited. Another important source from which the knowledge of Greek culture was derived was the numerous Syrian Christian monasteries which were scattered over the country.[12]

Under Al-Mamun's patronage, although the Muslim world was busy with military expeditions, they also applied their energy and knowledge to acquire knowledge and scientific research. Muslim scientists have cultivated and improved astronomy, geography, mathematics, medicine, and almost every branch of higher knowledge in the sciences and literature. Al-Mamun established libraries, laboratories, and observatories in Bagdad. Under Al-Mamun, they had discovered the equations of the second degree, and very soon after they developed the theory of quadratic equations and the binomial theorem. Not only algebra, geometry, and arithmetic, but optics and mechanics made remarkable progress in the hands of the Muslims.

Muslim scientists invented spherical trigonometry; they were the first to apply algebra to geometry, to introduce the tangent, and to substitute the sine for the arc in trigonometrical calculations. Their progress in mathematical geography was no less remarkable. The works of Ibn Haukal, of Makrizi, Al-Istakhri, Mas'udi, Al-Beruni, Al-Rumi, and Al-Idrisi, Kazwini, Ibn ul-Wardi, and Abu'l Feda, show what the Saracens attained in this department of science, called by them the Rasm ul-Arz. At a time when Europe firmly believed in the flatness of the earth and was ready to burn any foolhardy person who thought otherwise, the Arabs taught geography through globes.[13]

The physical sciences were as diligently cultivated. Chemistry, botany, geology, natural history, among others, occupied the attention and exercised the energies of the ablest men under Muslim rulers. Chemistry, as a science, is unquestionably the invention of the Muslims. Abu Musa Jabir Ibn Hayyan (the Geber of Christian writers) is the true father of modern chemistry. Jabir compiled a work of two thousand pages in which he inserted the problems of his master, Jaafar as-Sadik, which formed five hundred treatises.[14]

The science of medicine and the art of surgery, the best index of a nation's genius and a severe test of the intellectual spirit of a faith, were developed to the highest degree. Medicine had undoubtedly attained a high degree of excellence among the Greeks, but the Arabs carried it far beyond the stage on which their predecessors in the work of civilization had left it, and brought it close to the modern standard. They invented chemical pharmacy and were the first founders of those institutions which are now called dispensaries. They established in every city public hospitals, called Dar ush-Shifa, (the house of cure), and maintained them at the expense of the State.[15]

Abu Bakr Mohammad Ibn Zakaria al-Razi, who flourished at the beginning of the tenth century, was principal of the public hospitals at Rai, Jund-Shapur, and Bagdad. His treatises on smallpox and measles have been consulted by the physicians of all nations. He introduced the use of minoratives, invented the seton, and discovered the nerve of the larynx. He wrote two hundred medical works, some of which were published in Venice in 1510.[16]

Ali Ibn Abbas flourished fifty years later than Ar-Razi and corrected many of the errors of Hippocrates and Galen (Jalinus). He published a medical work, consisting of twenty volumes, on the theory and practice of medicine, which he dedicated to the Buyide Ameer Azud ud-Dowla. This work was translated into Latin in 1227 and printed at Lyons in 1523 by Michel Capella.[17]

Abul Kasim Khalaf Ibn Abbas, or Albucasis, was not only a physician but a surgeon of the first rank. He performed the most difficult surgical operations in his own and the obstetrical department. In operations on women, we are informed by him, in which considerations of delicacy intervened, the services of properly instructed women were secured. The ample description he has left of the surgical instruments employed in his time gives an idea of the development of surgery among the Arabs.[18]

Abu Ali Husain Ibn-Sina, or Avicenna, was unquestionably the most gifted man of his age; a Universalist in genius and encyclopaedic in his writings. He was also a philosopher, mathematician, astronomer, poet, and physician; he has left his influence impressed on two continents, and well deserves the title of Aristotle of the East. In spite of patristic jealousy, his philosophic ideas exercised an undisputed sway for several centuries in the schools of the East as well as in Europe. Ibn-Sina, a Persian-speaking scientist from the tenth century, also made important contributions to the study of the nature and philosophy of religious belief. He also found time to invent an early example of the micrometer, and his book *The Canon of Medicine* was taught to trainee doctors in universities in France and Italy from the twenth to the sixteenth century.

Avicenna is commonly known in Asia as "*The Shaikh par excellence*". He was born in the year 980 at a village called Afshanah in Transoxiana, of which place his father was the governor. He finished his medical studies in Bokhara at the age of eighteen, when he commenced an extraordinary political and philosophical career. His tenacity in refusing the liberal offers of Mahmud the Conqueror to join his service led to his expulsion from the Ghaznavid dominions. He quickly rose to the position of Vazir of Shams ud-Dowla, Ameer of Hamadan, and later Vazir of Ala ud-Dowla, Ameer of Isphahan, where he pursued his scientific and philosophical studies and wrote his great works, the Kanun and the Arjilza, which became the foundation of all medical knowledge.[19]

Abu Merwan Ibn Abdul Malik Ibn Zuhr, or Aven-Zoar, was one of the most distinguished physicians of his age. After finishing his medical and scientific studies, he entered the service of Yusuf bin Tashfin, the great Almoravide monarch of Africa, who covered the rising physician with honours and riches. Ibn Zuhr joined, like Albucasis, the practice of medicine with surgery. He was the first to conceive the idea of bronchotomy, with exact indications of the luxations and fractures, and he discovered several important maladies with their treatment. His son also followed in his father's steps and was the chief surgeon and physician of Yusuf bin Tushfin's army.[20]

The decline of the Abbasids was reflected in stagnation, and mental freedom was impaired. Though individual rulers continued to foster learning and scholars carried on their peaceful pursuits without difficulty, the leadership and encouragement of learning as an important part of imperial policy were lacking after the death of Al-Wathiq in 842-847. The so called growing linguistic superiority, religious orthodoxy, and spirit of intolerance, which developed in the later mediaeval period, were the main reasons for the decline in the scientific temperament. Above all, the rulers and Sardars in the later mediaeval age did a great disservice to scientific research; they were more concerned with religious practices and political affairs.

The Muslim theologians and intellectuals just forgot that the Holy Quran taught the one universal science and began to look the science as Hindu science, Christian science, Zionist science, Confucian science etc. However, the later Mediaeval scholars did not go beyond their forefathers' achievements and never gained the assistance and knowledge from the European progress in science and technology. The Muslim scholars were rather glad to be separated from modern science because they were of the view that through scientific research, Islam would be corrupted or be mixed with innovation.

The gap of acquiring knowledge and science was filled by the European and American intellectuals who were more practical in experimental observations and research. Moreover, they used Muslim knowledge and science as their basis of observation and scientific progress. There is no question that today, of all civilizations on this planet, science is weakest in the lands of Islam. Nowadays, many Muslims acknowledge and express profound regret at this fact, and at present, conferences are being held in many Muslim countries by specialists in the various branches of science in order to reveal their amazing achievements. The Muslim intellectuals are now needed to work more hard to change the course in the direction of true Islam for the good of the world at large.

Caliph Al-Mutwakkil (847-861) and Suppression of Mutazilite or Rationalists

The tenth Abbasid Caliph, Abu Al-Fadl Jafar Ibn Mohammad Al-Mutasim Billah, better known as Al Mutawakkil, had suppressed rationalism and scientific progress during his reign. At the time of his ancestors, the Rationalists, or Mutazilites, were the directing power of the state; they held the chief offices of trust; they were professors in colleges; superintendents of hospitals; directors of observatories; they were merchants; in fact, they represented the wisdom and the wealth of the empire; Rationalism was the dominating creed among the educated, the intellectual, and the influential classes of the community. However, the orthodox sect known as Sifatism was in force among the lower strata of society, and most of the Qazis, the preachers and lawyers of various degrees, were attached to it.

Mutazila[21] (*Ahl al-Adl wa al-Tawhid:* The People of Justice and Monotheism), rightly defined as a rationally oriented group of theologians, was an Islamic school of thought that believed in progress and the proper utilisation of mental faculty in terms of reason. They were ardent supporters of the progress and development of philosophy and science. They had the patronage of Caliphs Harun, Mamun, and Wathiq. Although Mutazilites or Rationalists were great learners, they were opposed to certain doctrines of Islam. They generally avoided theological questions and focused their energy on the science of nature and philosophy, with some influence from Greek philosophy. Without directly assailing the faith, they avoided all conflict with it so far as possible. However, upon the repeated enquiry of the orthodox party, the Mutazilites answered them in terms of the reason, which raised the controversy between the two parties. But it also must not be ignored that most of the philosophers like Ibn Sina, Ibn Rushd etc. were pure Muslims and believers in the Quran and Islam.

Mutazilites believe that the Holy Quran is created words of Allah rather than having always existed and thus being "uncreated" (everything except Allah is created by Him, even His own speech, Quran). They also believed that the Almighty Allah would not be visible in the Judgment Day. They denied the existence of the Attributes in God, and contested everything which could prejudice the dogma of the Divine Unity. To remove from God all ideas of injustice, they recognised man's entire freedom of action.

They taught that all the truths necessary for salvation belong to the domain of reason, and that they may be acquired solely by the light of reason, no less before than after Revelation, in such a wise that men, at all times and in all places, ought to possess these truths. As a result, a school of Mutakallims arose, who used Kalam in order to meet the arguments of the Mutazilites. Later on, all these arguments were systematized, and the Mutakallims devoted themselves to adducing reasons in defence of the faith as they conceived it.[22]

During the time of Caliph Mamun and Al-Wathiq, Mutazilites were bitterly opposed by the orthodox Islamic preacher Imam Ahmad Ibn Hanbal and supported by lower strata of society. He suffered greatly for the right to criticise science and rationalism. He was not even in favour of defending his faith with arguments. The great traditionist Ahmad Ibn Hanbal was imprisoned, flogged, and threatened with death when he refused to accept the doctrine of the createdness of the Quran. The movement for dogmatism spread in Mesopotamia, and a riot and bloodshed spread in the streets of Bagdad, which led to the arrest of Imam Hanbal, and the Imam died in captivity.[23]

Despite official support from the Abbasid Caliphs, the tide soon turned against the Mutazila when Caliph Al-Mutawakkil, the brother and successor of Al-Wathiq, ascended the throne. He was in alliance with the Sifatists, or orthodox section of society and proved a model as Caliph of the bigots, expelling the party of progress from their offices under government. The orthodox party, who now posed as martyrs, and the ignorant mob, with their febrile passions and violent prejudices, took up the cause of the oppressed, and the persecuted leaders became the leaders of the people.

Al-Ashari (873-936), a great Muslim Arab theologian, defended the fundamental Islamic belief that the Quran is a revealed book of Allah and that the religion of Islam must be based on it and the Prophet's Traditions. Al-Ashari also opposed the Mutazilites and held that man cannot create anything, the only creator is Allah; and further more man's power does not produce any effect on his actions. Allah creates in man a capacity for action which corresponds to the power and choice thus created. It follows that the action of the creature is created by God so far as initiative and creative power are concerned; but it is acquired by the creature, as it corresponds to his power and choice, with which he had been previously endowed.[24]

Under the Caliph Al-Mutawakkil, the study of nature began to be wholly neglected, and man took refuge in vague and futile doctrines, instead of conquering the forces of nature and harnessing them to the needs of man. The spirit of scientific investigation was replaced by credulity and superstition. The man's mind was turned from the scientific study of the phenomena of nature and the positive spirit of free inquiry applied to the solution of social problems. Those who loved free inquiry and advocated mental freedom found themselves ostracized. The slightest deviation from fixed prejudices was punished with severe. Philosophy was under a cloud, and many brilliant men, obstructed in their path, took refuge in mysticism. The colleges and universities were closed; literature, science, and philosophy were interdicted; and the Rationalists were hunted from

Bagdad. Al-Mutawakkil, at the same time, demolished the mausoleum of the Caliph Ali and Al-Hussain, the martyr of Karbala, and the site was ploughed and sown and nobody was permitted to visit the holy ground.[25]

Rise of Sunni Imamate and Decline of Science and Rationalism

The successors of the Prophet and the Muslim ummah followed the Quran as the primary source of law through its liberal and unscholastic interpretation. They also followed the precedents set in the Prophet's Traditions which they had helped form and to which they were a living witness. When the spirit of the Quran or Sunnah, public interest, and changes in time, place, intentions, environment, or social organisation required it, they resorted to new interpretations and innovative applications of the Quran. During this period, other sources of law, such as *Ijmaa* (consensus) and various nuances of *Ijtihad* (independent reasoning by Islamic expert), were also practiced.

These periods followed the mighty Muslim empire, and the rulers made educational development and scientific spirit rampant. The Abbasid Caliphs of Mansur and Mamun promoted the educational and scientific spirit, which led to the emergence of another thought of Islamic orthodoxy and Islamic revivalism. Some theologians issued the decree that the doctrine of the creation of the Quran is the only truth, and is binding upon all Muslims. Research of nature, science, and philosophy was considered unholy and an act of heresy.

In these very situations, a few sections of the population took charge of theological philosophy and propagated Islam according to their knowledge of research. However, Islam never granted to the Imam, or any leader of the Muslim people, any position but that of a protector and representative. Four scholars, Imam Abu Hanifa (699-767), Imam Malik (711-795), Imam Hanbal (780-855), and Imam Shafi'i (767-820), defined the Sunni Islamic philosophy in a new way and developed their own schools of thoughts, such as Hanafi, Maliki, Hanbali, and Shafi'i. Thus, four different systems of law and doctrine, more or less distinct from each other, thus established themselves in the Islamic world.

Malikism was practiced by a large proportion of Arabs in the Peninsula, Berbers, and the majority of Spanish Muslims; Shafi'ism was practiced by the middle and upper classes; and Hanafism was practiced by the more respectable sections of society in Mesopotamia, Syria, and Egypt. Hence, whilst rationalism ruled in the colleges and madrasahs, Hanafism held possession of the pulpits and Mahkamas. In its theological views, Hanafism inclined towards Sifatism; but it varied its opinions according to those of the rulers. In this period, Hanafism was remarkable for its flexibility.[26]

Hanafi School: It was during the eighth century that the earliest systematic efforts were made to lay the foundations of a science of Islamic law. The deliberations of a private 'legislative council' in Iraq, comprising Imam Abu Hanifa (Numan Ibn Thabit Ibn Zuta Ibn Marzuban 699-767) and his pupils, provided the basis for the establishment of the first widespread school of Islamic law, the Hanafi School. Two students of Abu Hanifa, Mohammad al-Shaibani and Abu Yusuf Yaqub al-Ansari, distinguished themselves in transmitting the Hanafi School. Abu Yusuf established the Hanafi School of Law in the Abbasid Empire in his capacity as Chief Justice-*cum*-Law Minister. This initiated the first move since the Rightly Guided Caliphs to bridge the gap between Islamic legal theory and legal practice.[27]

Jafari School: The other great jurist in Iraq contemporaneous with Abu Hanifa was Jafar al-Sadig (Jafar Ibn Mohammad Ibn Ali al-Sadiq, 702-765). He was the sixth Imam of the twelve, and Ismaili denominations of Shia Islam. He became the eponymous founder of the Imamiyyah, or Jafari School of law. His lectures were attended by Abu Hanifa and Malik Ibn Anas.[28]

Maliki School: Malik Ibn Anas (Malik Ibn Anas bin Malik bin Abi Amir bin Amr bin Al-Harith bin Ghayaman bin Khuthayn bin Amn bin Al-Harith al-Asbahi al-Madani, 711-795) became the founder of the Maliki School of Law, or the School of Tradition (Sunnah) in Madina. The Hanafi School in Iraq, called the School of Opinion (ra'y, ijtihad), made greater use of individual expert opinion and, in the beginning, tended towards building a logically perfect legal system based on pure reason and scholastic subtleties. The Traditionist school of Malik, flourishing in the milieu of the Prophet's living Sunnah in Medina, emphasised the real cases and temporal situations against the above tendency towards abstract thinking in law. However, in the beginning, the Maliki School tried to eternalise the decisions given in concrete cases in Medina.[29]

Shafi'i School: It was Al-Shafi'i (Abu Abdillah Mohammad Ibn Idris al-Shafi'i, 767-820), was another Arab Muslim scholar, author, and theologian. He tried to integrate and reconcile the Hanafi and Maliki schools, and became the founder of the Shafi'i School of law. Al Shafi'i was the first to compile the sources of Islamic law. His systematic reasoning and disciplined methodology in explaining the origins of Islamic jurisprudence left their mark on the finished structure of all the classical Islamic schools of law. Among the pupils of Al-Shafi'i were founders of new schools, such as Ahmad Ibn Hanbal, the founder of Hanbali school.[30]

Hanbali School: Ahmad Ibn Hanbal (Ahmad bin Mohammad bin Hanbal Abu Abd Allah al-Shaybani 780-855), was the founder of one of the great schools of Muslim law. He was the trust and strength of the orthodox and was rather more puritan than other Imams. He was against the pharisaical liberalism of Hanafism and disgusted both with the narrowness of Malikism and the common-place character of Shafi'ism. He adopted the extreme Sifatia views based on traditions and applied himself to framing a new system for the whole empire. He denounced Mutazila School and learning and science, and proclaimed a holy war against Rationalism. The populace, carried away by his eloquence or his vehemence, took up the cry; the Hanafi jurists, whose power materially depended on their influence over the ignorant masses, and who were jealous of the prominence of the scientists and philosophers in the Court of Harun and Mamun, made common cause with the new reformer.[31]

In the Islamic empire, more particularly in Bagdad, the men of the two philosophies, i.e. Islamic jururisprudence and upholders of reason began to fight against each other. The pulpits began to culminate in brimstone and fire against the upholders of reason and the advocates of philosophy and science. The streets of Bagdad became the scene of frequent rioting and bloodshed. Mu'tasim and Wathiq repressed the fanatical violence of the fiery puritans with some severity. The prime mover of the disturbances was put in prison, where he died in the odour of great sanctity; his body was followed to the grave by a crowd consisting of a hundred and forty thousand men and women. His system never took root among any large body of people, but mixing with Hanafism, it gave a new character to the doctrines of Abu Hanifa. Henceforth, Hanafism represents a mixture of the teachings of Abu Hanifa and of Ibn Hanbal.[32]

However, the greatest exponents of classical Islamic law, Ja'far al-Sadiq, Abu Hanifa, Malik, Al-Shafi'i, and Ibn Hanbal, never intended to found legal schools, much less to make their opinions valid for ever. These schools took their names posthumously. Most of the important differences between the legal schools did not outlive them. The students of one jurist often migrated to other provinces of the Islamic world-state, studied under jurists of another school, and a cross-fertilization of ideas took place. Thus Al-Shafi'i, a pupil of Malik in Medina, travelled in Iraq and studied with Al-Shaibani of the Hanafi School of Opinion. The appearance of nineteen major schools of Islamic jurisprudence during nearly 250 years, from about the seventh up to the tenth century, shows further the lack of intellectual rigidity and slavish imitation among these Muslims.

Mystical Philosophy of Islam or Sufism

After the crystallisation of the classical schools of Islamic law, independent inquiry and innovation among Muslim jurists gradually declined after the eleventh century. The extreme rationalism and intolerant attitudes of some Muslim jurists and philosophers created their own antithesis, a reactionary trend towards over-organisation, traditionalism, and social and legal rigidity. The rise of Muslim scholasticism drove some acute legal minds to gnostic Sufism, which is speculative and, as a form of free thought, an ally of rationalism.

Muslim mystic philosophy or Sufism may be regarded as the love of Supreme Beauty or the Union of the Beloved. Mysticism seems to have flourished between the eighth and fifteenth centuries in Europe and Asia. Muslim mystics and Sufis wielded considerable influence on the masses as well as the classes, and many of the leading Sufis were zealous missionaries. Sufism passed by easy gradations into mysticism, and around 820 these ascetics began to be known as Sufis.[33]

As in Islam, each person is potentially Allah's lieutenant on earth and everyone can attain perfection. So in the Middle Ages many Sufis attained perfection and preached human unity with the Almighty as the chief aim of mankind. To Muslim saints and Sufis, union with Allah is the ultimate end of existence, and a long line of Muslim Sufis, whose brilliant work has illumined Muslim faith, regard absorption in Allah as the goal of a Sufi's perfection. The great Persian poet and mystic saint Jalal ud-Din Rumi expressed the same idea in his inimitable verses:[34]

> "From the inorganic we developed into the vegetable kingdom,
> Dying from the vegetable we rose to animal,
> And leaving the animal we became men,
> Then what fear that death will lower us?
> The next transition will make us an angel,
> Then shall we rise from angels and merge in the nameless,
> All existence proclaims "Unto Him, shall we return".

In other words, the doctrine of Muslim mystics may be defined, "When a man becomes annihilated from his attributes, he attains to perfect subsistence; he is neither near nor far, neither stranger nor intimate, neither sober nor intoxicated, neither separated nor united, he has no name or sign or brand." This may be supplemented by a verse from the Quran: "Oh, thou soul which art at rest, return unto thy Lord, pleased and pleasing Him, enter thou among my servants, and enter thou my garden of felicity." We are the architects of our own fortune, and our future is shaped by our present life.[35]

The great lawyer, theologian, and mystic, Abu Hamid Ghazali (c.1056-1111), was the greatest representative of mysticism among the Muslims. He was a critic of the new sciences. He wrote a famous polemic, *The Incoherence of the Philosophers*, against the claims of scientists to be able to explain everything. The defender of orthodoxy, Al-Ghazali, made the knowledge stage open for a purer Islam based on the practice of the Prophet, not the fallible speculations of human philosophers and scientists. Imam Al-Ghazali was dissatisfied with the philosophical knowledge based on experience or reason or rationalism and had taken refuge in Sufism. However, he advocated the service of humanity through scientific means. At length, Imam Abu Hamid Ghazali wrote that acquiring knowledge of medicine and mathematics, indeed all beneficial knowledge, is a *Fard Kifayah* that requires a certain number of citizens to take up careers in such essential fields to serve mankind and advance the noble cause of bringing benefit to humanity.

Al-Ghazali felt obliged to provide a definitive Ash'arite response to the philosophers in a work entitled *Tahafut al-Falasifah* (The Incoherence of the Philosophers). The book argued that Mutazila's philosophical doctrines were either heretical or outright unbelief. Regardless of Ghazali's intentions, his work set the stage for the massive incorporation of arguments and concepts derived from or informed by philosophy. Al-Ghazali's influence served greatly to promote the diffusion of Sufism among the Eastern Muslims, and his idealistic philosophy was embraced by the greatest intellects of the Muslim East.[36]

Ibn Tufail (1110-1185) in the West, was another great theologian and representatives of mysticism among the Muslims. The poet Maulana Jalal ud-Din Rumi (1207-1275), whose *Masnavi* is one of the greatest mystic poems in the world, gives a comprehensive account of Sufism; Sanai (1080-1131), whom Jalal ud-Din himself has called his superior; Farid ud-Din Attar (1146-1221); Shams ud-Din Hafiz (1315-1390); the moralist Saadi (1210-1292); Khaqani (1120-1190); the romancer Nizami (1141-1209), all belonged to this school of Sufism.[37]

It must be admitted that Ghazali deals with the various schools of philosophy objectively, and after subjecting them to a critical analysis, rejects the doctrines of some as unbelief and of others as heresy. He began to study the writings of mystics and found that it was not through intellectual discipline but through feeling and emotion that he could understand Sufism. He had completed the circuit of thought and came back to the faith of his people after profound study and meditation. Having undergone intensely a variety of religious experiences, in his quest for the eternal he came to believe firmly in God, prophecy and in the last judgment.

Imam Ghazali's profound learning proved successful in incorporating mysticism into Islam, and henceforth Sufism was not only tolerated but encouraged by Muslim rulers.[38]

Imam Ghazali was not supposed to be the first preacher of 'inward light' in Islam, but intuitive knowledge of God is inherent in the Faith and communion with Him is the essential preliminary to true devotion; the *Meraj* of the Prophet meant the absolute communion of the Finite with the Infinite. Not only does God speak to the hearts of men and women, who in earnest sincerity seek divine help and guidance, but all knowledge is from the Supreme Intelligence; it comes to the Prophets by direct revelation, speak to the the hearts of His chosen few and to the others by intermediary.

Ghazali accepts the discoveries of science and adds that the most important facts and laws have been divinely revealed. He taught that the heavens and the earth are direct works of Allah, and were produced out of nothing by His Will; they are guarded by His Will and one day will pass away at His Command. Ghazali thus attempts the harmony of science with religion, and declares that behind the order of nature lies the omnipotent Allah, who says "Be" and the Universe comes into being. Ghazali says that we cannot gain positive knowledge through reason alone, but it may also be gained in books sent down to prophets or in the personal revelations of Allah's saints.[39]

Imam Ghazali rescued Islam from barren and stereotyped phrases and dogmas and brought them into living contact with the Holy Quran and the traditions. He destroyed the dead wood of Muslim scholasticism and emphasised the element of fear in his doctrines. He had felt the horrors of hell, and he made other people feel them. Moreover, it was through Ghazali's powerful personality and tremendous expositions of mysticism that Sufism found an assured place in the bosom of Islam.[40]

Jalal ud-Din Rumi's conception of free will is by no means free from ambiguity, as he admits that man is free, to a certain extent, but this freedom is subject to the Divine Will. He states that perfect love is perfect freedom, and this leads to the union of the human will with the Divine Will. For example, Jalal ud-Din Rumi holds with other Sufis that intellectual discipline and metaphysical arguments are inadequate means for knowing Allah. We can acquire knowledge of Allah through spiritual discipline, contemplation, purification of the heart, and devotion. According to Rumi, man must go through a number of intermediate stages and forms until he attains the perfection necessary for the soul, and he continues developing it until he has attained maximum development of his faculties. Rumi traces the progress of man from inorganic matter to the higher stages until he reaches the position occupied by man in the universe.[41]

In general, it is to be said that in Sufism or in mystic philosophy, there is no place for reason or rationalism. Almost all its philosophy depends upon the connectivity of the heart with the Almighty Allah, or in other words, the union of the human soul and God. The rise of mystic philosophy and its dominance over the centuries led to the decline of science and the suppression of reason and rationalism. Just as mysticism destroyed the foundations of rationalism and empirical philosophy, its influence on scientific enquiry proved disastrous.

NOTES

1. Prayer has been an obligatory duty since the Islamic faith emerged in Makkah. The early prayer was performed individually twice a day at certain times: early morning before sunrise (*Fajr*) and late afternoon before sunset (*Asr*). The two prayers consisted of the repetition of two units called *Rakat*, consists of prescribed actions and words. In the tenth year of the Islamic propagation (620), the Prophet took a long journey from Makkah to Jerusalem, ascended to heaven, and returned to Makkah in a single night. The event was marked as a Night Journey (*Isra* and *Meraj*). As a result, he was ordered to lead prayer five times a day. He then instructed the faithful followers the details of prayer; evening (*Maghrib*), afternoon (*Dhuhr*) and night (*Isa*) prayers were ordered to be performed in addition to the two prayers conducted before the event. There was no clear evidence about when Muslim started to perform the five time prayers—before or after the migration to Madinah, but group prayers were clearly introduced in the first Islamic mosque built in the first year of migration to Madinah.
2. Fasting is observed in the month of Ramadan month in which the Holy Quran was sent down as a guide to mankind. The first verse regarding fasting in the month of Ramadan was revealed to Prophet Muhammad in February 624, or in the month of Shawwal in the second year of *Hijrat*, in Madinah. Allah revealed in the Holy Quran, Surah Baqra (2:185), that, "O believers! Fasting is prescribed for you, as it was for those before you, so perhaps you will become mindful (of Allah)". This verse was obligatory for everyone who had reached puberty, except sane, ill, and elderly people. Prophet Mohammad and some Muslims fasted some days, in particular months in Makkah prior to the Quran's Ramadan verses, fasting 30 or sometimes 29 days straight without any interruption. A short time prior to the revelation of Ramadan verses, Muslims also changed their prayer direction (*Kiblah*) from Jerusalem (*Quds* in the Quran) to Makkah's Kaabah, the cubic structure built by Prophet Abraham to pray to the One Allah. All these changes happened after the Prophet's followers established a strong base in Madinah. Also, the first Ramadan remarkably coincided with the first crucial military engagement, the Battle of Badr, between Madinah based Muslims and Makkah led paganists.

3. Kabir, Humayun (1955). *Science, Democracy and Islam*, London, pp. 19-20. In religions other than Islam, priests enjoyed power without responsibility. If power corrupts, power without responsibility corrupts even more.
4. Bhargava, *op. cit.*, p. 92.
5. *Ibid.*, pp. 92-93.
6. *Ibid.*, p. 93.
7. Iqbal, Muzaffar, *op. cit.*, pp. 20-21.
8. For Aristotle, the cosmos was a gigantic spherical plenum that had neither a beginning nor would ever have an end. Aristotle rejected the possibility that other worlds might exist beyond our own. Within the cosmos, Aristotle distinguished two major divisions, the celestial regions and the terrestrial. The terrestrial region, just below the concave surface of the moon, was a region of constant change and transformation. It consisted of four elements—earth, water, air, and fire. In the terrestrial region, bodies were always coming into being and were always passing away because their elements eventually dissociated to combine with other elements and form new compound bodies. In the upper atmosphere of the terrestrial region Aristotle assumed that, comets, shooting stars, and other similar changeable phenomena occurred. Aristotle attributed to celestial regions those suffer no change except that of place. It did not come into existence and will never pass away. He regarded the planets and the stars as composed of ether and thus also undergoing only observable changes of place. Although not created by God, Aristotle's cosmos was a rationally constructed cosmos in which a certain order prevailed. (Iqbal, Muzaffar, *op. cit.*, pp. 98-99.)
9. There is a famous story that one night the philosopher Aristotle appeared to the Caliph Mamun in a dream, telling him to seek what was good according to reason. This dream was one of the reasons that the Caliph initiated a project to translate Greek scientific literature into Arabic. The Caliph then wrote to the Byzantine emperor asking for manuscripts to translate. Although at first reluctant, the emperor eventually complied, and a delegation was dispatched from Bagdad to acquire the manuscripts. After that, books on philosophy and other ancient sciences became plentiful in the Islamic empire. [Walbridge, John (2011). *God and Logic in Islam: The Caliphate of Reason*, New York. p. 54.]
10. Bhargava, *op. cit.*, pp. 185-186.
11. *Ibid.*
12. *Ibid.*
13. Ameer Ali, *op. cit.*, p. 384.
14. *Ibid.*
15. *Ibid.*, p. 385.
16. *Ibid.*
17. *Ibid.*
18. *Ibid.*, p. 386.
19. *Ibid.*, pp. 386-87.

20. *Ibid.*, p. 386.
21. Mutazila, originated at Basra during eighth century, was an Islamic school of thought appeared in the early Islamic history in the dispute over Ali's leadership over Muslim community after the death of the third Caliph Othman. Those who would neither sanction nor condemn Ali and took a middle position were termed Mutazila. The later Mutazila School developed an Islamic type of rationalism, partly influenced by ancient Greek philosophy. Their notorious doctrine, that the Quran was created, against a widely accepted claim that elevated the Quran to a degree of un-createdness (similar to Allah) and its origin in a "heavenly tablet" (*Al-lawh al-Mahfuz*). This ideology invited the opposition of orthodox Sunni Muslims in the form of Ashari, Matuiridi, and the traditional school of theology. They upheld the doctrine of the created Quran as they insisted on the eternal nature of God. They also asserted the principle of the necessary justice of Allah: He wills and does only that which is good, and thus they limit the scope of Allah's actions.
22. Prophet Mohammad (pbuh) had said, "One day ye shall see your Lord as you saw the full moon at the Battle of Badr," and these words, which the orthodox took literally, were for Mutazilite an ever new stumbling block. They, therefore, explained them away by saying that man, after his death, would know Allah by the eyes of the spirit, that is to say, by the reason. Mutazilites were perhaps overconfident and full of pride and did not even hesitate to frankly declare that it was not impossible to write something as good as, or even better than, the Quran. They could not admit the miracles related in the Quran, and so denied that the sea was dried up to yield a passage for the Israelites led by Moses, that Moses' rod was charged into a serpent; and that Jesus raised the dead to life. However, their idea of Allah was more exalted than that of the orthodox. (Bhargava, *op. cit.*, pp. 191-92 & 201.)
23. Ameer Ali, *op. cit.*, p. 440. [Ibn Hanbal (780-855) was a *Hadith* scholar, the eponymous founder of the Hanbali legal school, and one of the great figures of textual literalism in Islam. A vigorous advocate of the authority of the *Hadith* against legal rationalism, he also opposed the Mu'tazila in theology, in particular their doctrine of the createdness of the Quran.]
24. Bhargava, *op. cit.*, p. 202.
25. *Ibid.*, pp. 186 & 195.
26. Ameer Ali, *op. cit.*, p. 438.
27. Husaini, S. Wagar Ahmed (1980). *Islamic Environmental Systems Engineering*, London. p. 17.
28. *Ibid.*
29. *Ibid.*
30. *Ibid.*, p. 18.
31. Ameer Ali, *op. cit.*, p. 439.
32. *Ibid.*

33. The Muslim mysticism or Sufism may be regarded as the love of Supreme beauty. The term Sufi derives from the woollen garment, the word "suf" meaning wool, and the term was applied to a person who had renounced the world and become an ascetic, which henceforward meant that he had become a Sufi. The term Sufi has no connection either with the *Ahl-us-Suffa*, the religious men who were want to sit and sleep outside the Prophet's mosque and receive their food daily from him, nor with the *Ikhwan-us-Safa*, the Brethren of Purity. In later periods, a number of Sufi orders were established in Iran, Bagdad, Samarkand, and India. Their number was not less than 175, but there were some who had a profound effect on the community through their saintly lives and missionary fervour. The most famous were: (1) Qadriya Order founded by Abdu Al-Qadir (1078-1166) of Gilan, North Iran. (2) Suhrawardiya Order founded by Diya Al-Din Abu Najin As-Suhrawardi (1097-1168). (3) Chishti Order was founded in Chisht in Heart by Moin Al-Din Chishti (1143-1236). (4) Naqshabandi Order founded by Khwaja Baha'uddin Mohammad (1318-1389). (Bhargava, *op. cit.*, pp. 221-22.)
34. Bhargava, *op. cit.*, p. 5.
35. *Ibid.*
36. Walbridge, *op. cit.*, p. 52.
37. Ameer Ali, *op. cit.*, pp. 456-57.
38. Bhargava, *op. cit.*, p. 216.
39. *Ibid.*, p. 217.
40. *Ibid.*, p. 218.
41. *Ibid.*, pp. 223-24.

6
Role of Muslim Scholars in Development of Science

The origins of Islamic sciences can be traced partly to Sumer, Babylon, Egypt, Greece, Persia, and India's scientific heritage; partly to inspiration derived from the Quran and the Prophetic words (*Hadith*) of the Prophet Mohammad (pbuh); and partly to the intellectual and creative genius of Muslim scientists, thinkers, and philosophers during those extraordinary many hundred years of Islamic history. It would seem that we need a satisfactory explanation to make sense of the development of Islamic science and the intellectual roots of Islamic civilization.

In the seventh century and afterwards, the Muslims conquered all of the Arabian Peninsula from Al-Hamrah in Spain to the borders of China. For centuries after the fall of ancient Rome, scientific progress in Western Europe slowed almost to a standstill. Muslim scholars commenced to acquire Greek expositions and established their study and conversion into Arabic a few centuries after the Hijrah. They significantly examined, assembled, approved, and incremented the science and philosophy of the Greeks. Following this period began what is known as the Golden Age of Islam. It is that we discovered many Islamic scientists who left behind hundreds and thousands of books on several branches of science.

In the developing Muslim world, however, a Golden Age of discovery flourished from the seventh century until the seventeenth century. During this period in the Muslim world, the Islamic scholars of various beliefs and cultures built and improved upon the knowledge of Greek philosophy and science, Persian literature, Indian medicine and mathematics, and certain aspects of Egyptian and Babylonian science of which even the Greeks had been unaware. During this period, there was great emphasis on the pursuit of knowledge. Muslims began to believe that Islam was not merely a set of commandments and rituals but a complete way of life, encompassing all domains of knowledge and human activity. They thought that all realms of knowledge were branches of the same tree.

Following the Islamic thought of seeking knowledge, the Islamic philosophers studied the Greek and Roman texts. The Islamic scholars found themselves in possession of the dazzling intellectual treasures of these ancient Greek and Roman civilizations. The pre-Islamic sciences came to be known as the *Ulum Al-Awa'il* (knowledge of antiquity). Indeed, the *Ulum Al-Awa'il* was a vast storehouse of intellectual treasures. It was really a challenge to assimilate these elements of secular learning into the Islamic faith, but they were taken up.[1]

The Muslims believed that they were commanded by Allah in the Holy Quran and by the Prophet Mohammad (pbuh) to study, seek knowledge, learn and benefit from others. It was this belief that inspired the Muslims to great heights in the sciences, medicine, mathematics, astronomy, chemistry, philosophy, art, and architecture. The mathematicians, astronomers, chemists, physicians, architects, engineers, economists, sociologists, artists, artisans, historians, geographers, educators etc. of the Muslim Golden Age made beneficial contributions to society and humanity with open-mindedness and, in many instances, in collaboration with people of other faiths, cultures, and backgrounds.

At this time, Islam was not just a set of religious beliefs, but a set of ideas, ethics, and ideals encompassing all aspects of human life, including science, literature, and philosophy. Science entered the Islamic world during the classical Abbasid period, but a few examples could be seen even before. During these Islamic heydays, Muslim scholars, such as mathematicians, astronomers, chemists, physicians, architects, engineers, economists, sociologists, artists, artisans, historians, geographers, and educators, made beneficial contributions to society and humanity. They did so with open-mindedness and, in many instances, in collaboration with people of other faiths, cultures, and backgrounds.

As for the science in Islamic civilization in first hundred years the original sources are not sufficient to declare that science had entered the Islamic society. Moreover, during these days Arabs were busy in their conquests and expansion of their territory including the two civil wars.[2] Some traces are found that some alchemical, astrological, and medical works were translated on the initiative of the Umayyad prince Khalid bin Yazid (d. 704), who had turned to the study of alchemy when his claims to the Caliphate were thwarted. By the second century of Islamic civilization, however, the enterprise of science in Islam was already well established, with definable branches and scientists of high calibre working in disciplines such as cosmology, geography, astronomy, and alchemy.

Thus, until we discover new manuscripts and other primary sources, the story of the emergence of science in Islamic civilization has to remain tentatively dated around 777, the year in which Jabir bin Hayyan, one of the most accomplished Muslim scientists of this early period, is said to have died. It was definitely the Abbasid period (750-1250), and during these times, Islam was in a state of vigorous territorial and commercial expansion. Islamic society was vibrant and dynamic, and the cultivation of the arts and sciences formally began in these times.

The geographical expansion of Islam within its first century was accompanied by a social revolution that reconfigured the social, cultural, and intellectual climate of the old world. The same social revolution provided an opportunity for Islamic civilization to receive a very large amount of scientific material from Greek, Persian, and Indian sources. However, prior to the arrival of Greek, Indian, and Persian scientific materials, there was an indigenous scientific tradition ready and able to comprehend and receive this material. This infusion was not a random process; rather, it was an organised and sustained effort spread over three centuries, involving thousands of scientists, scholars, translators, patrons, books, instruments, and rare manuscripts. Islam's contribution to the scientific enterprise was complex and rich, and it spanned over three continents and nearly a millennium of time.

Muslim scholars studied nature in the context of the Quran. The Quran depicted the relationship between nature and man, and this inspired the Muslim scholars to study natural phenomena in order to understand Allah. Moreover, the Islamic scholars borrowed knowledge from the entire world by translating Greek, Persian, Egyptian, Babylonian, and Indian works and gained a lot for the further progress of science through research and observations. The Islamic scientific traditions that flourished during the Islamic heydays continued to inspire European scholars for their scientific development. Astronomy flourished at the beginning of science in the Islamic tradition, went through a static period, and then burst once more into the forefront. Alchemy had initial energy and then remained unchanged for several centuries. Mathematics developed steadily throughout the eight hundred years, as did geography and geology.

Translation Movement

The translation movement was started by the Abbasid rulers to gain knowledge in various fields of science and philosophy. Many of the achievements of the Islamic Arabic Golden Age were based on previous initiatives taken by the ancient Egyptians, Hebrews, Persians, Greeks, and Romans. Hence, translators were invited to Bagdad, where scientists and researchers studied the past and created the future. The translation movement began slowly in the Caliphate of Al-Mahdi (775-786) and Harun al-Rashid (786-809) but really got underway in the time of Al-Mamun (813-833). Soon, ancient manuscripts were flooding into Bagdad (and also Basra), ready to meet the demand. Most were Greek, but they also came from Persia, from India, and maybe even China too. The result of their work was impressive progress in all sectors of science. The rulers of Islamic Spain, in an attempt to surpass Bagdad, recruited scholars who made contributions of paramount importance to science, medicine, technology, philosophy, and art.

Al-Mamun and other members of the elite sent out missions to find manuscripts. One possibly apocryphal story tells of a mission to Byzantium, where after some persistence, the manuscript hunters were told that they might find Greek manuscripts locked away in an old church. When they finally gained entry, they found many of the great pieces of Greek scholarship in a fragile state, covered in dust and cobwebs or even rotting away with mildew. Some of the manuscripts for translation were recovered by official missions in this way. Others were, no doubt, brought to Bagdad by those eager to cash in on the booming market.

The main scientific works of men such as Hippocrates, Aristotle, Theophrastus, Euclid, Ptolemy, Dioscorides, Galen, and many others were rendered into precise Arabic thanks to masters of translation such as Hunayn Ibn Ishaq and the concerted effort of Caliphs, princes, and vazirs. Many of the translators were not Arabs, nor even Muslims, but from faiths and languages brought in by the huge expansion of the Islamic realm. Many were native Greek speakers from the old Byzantine Empire. Many, too, were Christian scholars who spoke and wrote Syriac, a form of Aramaic, the original language of the Bible and adopted by many eastern Christians between the fourth and eighth centuries. Many of the texts were actually translated first into Syriac and only then into Arabic in a two-step process.

Through this movement, Arabic became the most important scientific language in the world for many centuries, and the ground was prepared for the rapid growth of the Islamic sciences, properly speaking. The translations provided the *materia prima* upon which the Muslim mind pondered and which it moulded into the substructure for the Islamic sciences that soon came into being as a body of knowledge at once distinct and related to the age-old intellectual traditions which it had inherited and adopted as its own thanks to the synthesizing and integrating power of the Islamic revelation.

The Muslim rulers, Muslim officials, and scholars made a concerted effort to commission Arabic translations of the major works of foreign science and philosophy, especially Greek, Persian, and Indian science and philosophy. There were also translations from Middle Persian, mainly works on astronomy, science and medicine, astrology, and practical and political wisdom, some of them originally written in Sanskrit. But Greek science required Greek philosophy to be understood properly, so a great many philosophical texts, including virtually all the works of Aristotle and a very large selection of commentaries, were translated.

The Indian scriptures and manuscripts were also very helpful to Muslim scientists. The Indian scientific tradition, especially in mathematics (including astronomy) and medicine, reached Islam not only through Sassanid Persia but also thanks to a number of Indian men of learning who were invited to Bagdad and other Islamic intellectual centres. Of course, the Indian sciences entered into the Islamic world again through the writings of Al-Biruni in the eleventh century and soon after through the numerous works of Amir Khusraw. But as far as the genesis of Islamic science is concerned, it was mostly as a result of the translation of certain basic texts of mathematics and astronomy, such as the *Brahmasphutasiddhanta* of Brahmagupta and a few medical works, particularly concerning drugs and poisons, that Indian science had a visible effect upon Islamic science and became one of the notable elements that contributed to its birth.[3]

The scholars from Bagdad did not simply take the works popular among the Byzantine Greeks at that time; philosophy and science were at a very low point in Byzantium during that century. Instead, they often went back to classics that had passed out of common circulation—translating, for example, Ptolemy's *Almagest* instead of simpler works that were widely read in the ninth century. In fact, there is reason to think that Islamic demand for these works was a major factor in bringing them back into circulation in Byzantium and thus assuring the survival of the Greek originals. It is most important that they took pagan philosophical works instead of Christian ones.[4]

From about the middle of the eighth to the middle of the eleventh century, a systematic, elaborate, sustained, and well-organized translation movement brought almost all philosophical and scientific books into Arabic. Under this movement, the Greek scientific tradition was translated into Arabic over a period of three centuries. The translation movement did not solely bring Greek science but also brought Indian and Persian scientific data and theories into Arabic. This process involved numerous influential people, who had diverse temperaments, racial and intellectual backgrounds, and reasons for their involvement in this task. The translation movement by Islamic scholars led to the development of science and natural history in Islamic civilization. This translation movement resulted in golden age of the Islamic civilization and history.

These translations planted the seeds of what became a major intellectual tradition in the next two centuries. It was a movement that enriched the Arabic language by compelling its philologists to coin new technical terms; compelled the best minds of the time to find ways to accommodate, discard, or transform theories and ideas that contradicted their religious beliefs; brought a massive amount of scientific and philosophical data into Islamic civilization; and produced tensions and conflicts within the Islamic intellectual tradition that were, in the end, greatly beneficial.[5]

The Translation movement had political, cultural, intellectual, and religious motivations. It came into existence owing to certain internal needs of the Islamic polity of the time, and, once in existence, it produced enormous creative energy in an intellectual climate already filled with curiosity, ready to use whatever it could for its new ventures. Although translation activity had already begun during the pre-Abbasid period, it was the Abbasids who provided the resources for a sustained and systematic process of translation of scientific texts into Arabic. The translation movement became more organized and received financial and administrative impetus after the founding of Bagdad by Al-Mansur.[6]

The Muslims did not just take over Greek science and translate it into Arabic and preserve its Greek character. It was totally transformed into a part and parcel of the Islamic intellectual citadel. It is true that the particular descriptions might have been taken from the works of Aristotle or a particular formula from Euclid's Elements, but the whole science is totally integrated into the Islamic point of view.

Massive amounts of Greek works translated into Arabic have given the false impression that Islamic scientific tradition arose as a result of the translation movement, when in fact Islamic scientific tradition predated the translation movement. Astronomy, alchemy, medicine, and mathematics were already established fields of study before any major translations were made from Greek, Persian, or Indian sources.

The scope of this translation movement can be judged from the range of subjects covered, which included the entire Aristotelian philosophy, alchemy, mathematics, astronomy, astrology, geometry, zoology, physics, botany, health sciences, pharmacology, and veterinary science. All these translations were done to enrich the tradition, not to create it, as some scholars have claimed. In fact, translated material went through constant and detailed examination and verification and was accepted or rejected on the basis of experimental tests and observations. This process of scrutiny started as early as the ninth century, that is to say, almost contemporaneously with the translation movement.

Muslims gained access to the Greek medical knowledge of Hippocrates, Dioscorides, and Galen (Jalinus) through the translations of their works in the seventh and eighth centuries. The tradition of the production of astronomical tables, for instance, may have been inspired by the Ptolemaic Handy Tables tradition, but the tables produced by Muslim astronomers were not merely an Arabic reproduction of Ptolemaic tables; they were the result of astronomical observations that began as early as the first half of the ninth century with the expressed purpose of "updating the Zijes, inspired by the Handy Tables". These initiatives by Muslims could be seen in the different aspects of the healing arts that were developed.[7]

Three distinct phases can be identified in this movement. The first began before the middle of the second century of Islam, during the reign of Al-Mansur. Major translators of this first phase were Ibn al-Muqaffa; his son, Ibn Na'ima; Theodore Abu Qurra, a disciple of John of Damascus who held a secretarial post under the Umayyad Caliphs; Thabit Ibn Qurrah, a Sabian mathematician; Eustathius, who along with Theodore Abu Qurra translated for Al-Kindi; and Ibn al-Bitriq, who was a member of the circle of the Caliph Al-Mamun. Al-Mamun's accession marks the beginning of the second phase of the translation movement. Many new translators were involved during the second phase of the translation movement, which was led by Hunayn Ibn Ishaq. These translators refined many translations from the first phase and extended the range of material being translated. Under the third Abbasid Caliph Al-Mahdi, the important translators were Nestorian patriarch Timothy I and Abu Nuh. During the third phase, beginning with the dawn of the tenth century and ending around 1020, of the translation movement, further refinement of the translated material took place and commentaries started to appear. The most important translators of this phase were Abu Uthman al-Dimashqi, Yahya Ibn Adi etc. The three-hundred-year-old translation movement had come to an end by the middle of the eleventh century. [8]

The greatest of the Muslim translators, Hunayn Ibn Ishaq, trained and led a number of translators, including his son Ishaq, who were responsible for translating a sizeable body of Greek works in medicine, philosophy, and mathematics into Arabic. Thabit Ibn Qurrah was another master of translation, who comprised a pagan cult deeply influenced by astrology and Pythagorean mysticism. Other major translators were Abu Bishr Matta and

Yahya bin Adi etc. This first stage of Islamic scientific development was essentially an assimilation of imported knowledge, and Muslim scholars had only a secondary role to play as translators.[9]

The Arab scholars not only translated the Greek and Roman texts but also wrote the commentary on those texts and produced a critical attitude towards the translated texts, which spurred a movement for their revision, both at the level of experiment and theory. Abu Bakr Zakaria Al-Razi's yet to be published *Kitab Al-Shukuk Ala Jalinus* (Doubts Concerning Galen), and Ibn Al-Haytham's *Al-Shukuk Ala Batlamyus* (Dubitations in Ptolemaeum) are excellent examples of this kind of literature.[10]

The extent of the social, political, and financial patronage this movement received can be gleaned from the social classes that supported it, which included Caliphs, princes, merchants, scholars, scientists, civil servants, and military leaders. Indeed, the court of the Caliph and the homes of the nobility received sages and scholars of all creeds as dignitaries. They were venerated and respected, not simply tolerated. Nurtured by this environment of liberalism and religious open-mindedness, science quickly took root in the lands of Islam. The study of this translation movement has yielded many valuable texts that have enhanced our understanding of the role of Greek science and philosophy in the making of the Islamic scientific tradition.

Progress of Science

The progress of science is based on several factors. Muslims following the guidelines of the Prophet studied and searched for knowledge. The Quran is clear: "The scholar's ink is more sacred than the blood of martyrs", while the Prophet promoted medical research, preaching that "For every disease, Allah has given a cure." Moreover, communication became easier because the Muslim Empire united extensive geographic areas. Scholars travelled to teach or share ideas. Furthermore, the Arabic language became a unifying factor. Translations from Greek, Latin, and Chinese into Arabic were innumerable, thus removing language barriers for scholars. During the same period, Arabs learned from the Chinese how to produce paper, and books became more readily available. Libraries were established in Cairo, Aleppo, Bagdad, and urban centres in Iran, Central Asia, and Spain, while bookshops with thousands of titles opened in several cities. Finally, the House of Wisdom, an academic institution serving as a university, was established in Bagdad in 1004.

Muslim science lasted for nearly six centuries, and this is longer than Greek, mediaeval Christian, or even modern science has lasted. However, the phase of Islamic civilization 1000 to 1250 was most crucial in terms of the development of science and technology in the Islamic world. Until this phase, the translation effort had been almost completed, and science in Islamic civilization had now entered into its second phase of development. Arabic, not Greek, has now become the vehicle of intellectual thought. In contrast to the earlier period, most scholars in the lands of Islam were now Muslims. Most of these scientists were interested in more than one branch of science, as was usual at that time. The highest concentration of scientific activity at this early stage is, however, in mathematics, astronomy, alchemy, natural history, and medicine.

In this phase of Islamic civilization, the Muslim community produced scientists such as Ibn al-Haytham (965-1039), Al-Biruni (973-1051), Umar Khayyam (1038-1123), and Nasir-ud-Din Tusi (1201-1274). The works of these scientists were the major source of research for European scientists during the Renaissance period. Roger Bacon, much to the displeasure of the Christian Church, began his experiments based on Ibn al-Haytham's treatise on optics. The Latin translation of Ibn Sina's Canon of Medicine was taught for centuries in Western universities. Ibn Rushd was destined to become the first philosopher of the Reformation.[11]

Here it is a question that how the theologians supported the scientific research, reason and free enquiry. Actually the orthodox theologians or Jabria orthodoxy always discouraged the free scientific enquiry and research but a school or movement Mu'tazila (Dissenters) started in the early eighth century by Wasil Ibn Ata, an ex-student of Imam Hasan al-Basri. The Mutazila sought a reconciliation of faith with reason. This synthesis of Muslim theology with Greek logic gave birth to a theological science, *Ilm ul-Kalam*. In their quest for rational and philosophical underpinnings of the Islamic faith, the Mu'tazilites advanced arguments based on ethics and reason, even though they naturally supported their positions using Quranic quotations. They argued that God has given people the power to make free, unfettered choices.[12]

It spread through the courts of the nobility in the Empire, into Spain, and thence into Andalusia. Caliph Mansur encouraged it without publicly committing to it, but Mutazilism (a liberal Muslim sect) was actually imposed as state doctrine during the reigns of Al-Mamun and Al-Mutassim. There cannot be any doubt that Al-Mamun was the greatest patron of philosophy and science in the entire history of Islam. His creation of the *Bait al-Hikmah* (House of Wisdom) was an important milestone. This immense scientific academy was an unrivalled centre for the study of humanities and sciences, where the greatest collection of worldly knowledge was accumulated and developed.[13]

In order to stock this official institute and library for research and translation, Mamun sent emissaries as far as Byzantium to seek out and purchase scientific and philosophical works. The doctrine of rationalism was preached in mosques and madrasahs, and became the distinguishing mark of the educated. The influential and intellectual classes of society, princes, courtiers, *qazis*, professors, doctors, and traders, accepted it as their creed. Phenomenal progress in the secular sciences occurred under Mutazilite rulers, and most of the great Islamic scholars and scientists either openly declared their allegiance to rationalism or were heavily influenced by it. Through the joint efforts of the rulers and intellectuals, the investigation and learning of science flourished for five centuries in Islamic civilization. The scholars like Al-Khwarizmi, Al-Kindi, Ibn Sina, Umar Khayyam, Ibn al-Haytham, Ibn Rushd, Ibn Khaldun, and many others tried their level best to enrich their society and curriculum with scientific learning and research.

Abu Musa Jabir Ibn Hayyan al-Azdi al-Tusi al-Sufi (721-815)

Jabir Ibn Hayyan, known in the West as Geber, was a Muslim alchemist known as the father of Arabic chemistry. He is reported to have studied under Imam Ja'far Sadiq and the Ummayed prince, Khalid Ibn Yazid. After completing his studies he finally went to Bagdad, where he first became known as an alchemist at the court of Harun al-Rashid and Mamun al-Rashid.

He introduced experimental investigation into alchemy, which rapidly changed its character into modern chemistry. His contributions of fundamental importance to chemistry include perfection of scientific techniques such as crystalization, distillation, calcination, sublimation, and evaporation and the development of several instruments for the same. Perhaps Jabir's major practical achievement was the discovery of mineral and other acids, which he prepared for the first time in his alembic (Anbique).[14]

He is the supposed author of an enormous number and variety of works in Arabic, often called the Jabirian corpus. The Jabirian corpus is so extensive and varied that some scholars have expressed doubts about its authorship by a single person. The scope of the corpus is vast and diverse, covering a wide range of topics, including alchemy, cosmology, numerology, astrology, medicine, magic, music, pharmacology, mysticism, and philosophy. There are about 600 Arabic works attributed to Jabir Ibn Hayyan that are known by name, approximately 215 of which are still extant today. Though some of these are full-length works, most of them are relatively short treatises and belong to larger collections in which they function rather more like chapters. When the individual chapters of some full-length works are counted as separate treatises too, the total length of the corpus may be estimated at about 3000 treatises/chapters.[15]

His achievements in this field include the following: preparation of various metals; development of steel; dyeing of cloth and tanning of leather; varnishing of water-proof cloth, use of manganese dioxide in glass-making; prevention of rusting; letterring in gold; identification of paints, greases etc. During the course of these practical endeavours, he also developed *aqua regia* to dissolve gold. The alembic is his great invention, which made easy and systematic the process of distillation. Jabir laid great stress on experimentation and accuracy in his work.[16]

In his works, Jabir described the preparation of many chemical substances: the sulphide of mercury, oxides, and arsenic compounds. Jabir Ibn Hayyan discovered vitally important acids like sulphuric, nitric, and nitromuriatic acid etc. Jabir Ibn Hayyan perfected the distillation process using the alembic still, which is still used today. By his methods, the Muslim world produced rose water, essential oils, and pure alcohol for medical use.[17]

His various breakthroughs, e.g., preparation of acids for the first time, notably nitric, hydrochloric, citric, and tartaric acids, and emphasis on systematic experimentation, are outstanding, and it is on the basis of such work that he can justly be regarded as the father of modern chemistry. In the words of Max Mayerhaff, the development of chemistry in Europe can be traced directly to Jabir Ibn Haiyan. Of Jabir's many books, among the most influential in the West have been *Kitab al-Kimya,* translated in 1144 by Robert of Chester as "The Book of the Composition of Alchemy," and *Kitab al-Sab'een,* which Gerard of Cremona translated in twelfth century. In chemistry, the works of Jabir Ibn Hayyan formed the basis of modern science.[18]

Mohammad bin Musa al-Khwarizmi (780-850)

Mohammad bin Musa al-Khwarizmi was born in Kath, part of Khwarizm. He was summoned to Bagdad by Caliph Al-Mamun and appointed court astronomer. He was one of the earliest scientific descriptive geographers, and a highly talented mathematician. His contributions to arithmetic, algebra, geography, and astronomy were to play an important role in the subsequent development of these sciences, both in the Muslim world as well as in Europe. His work on algebra was outstanding, as he not only initiated the subject in a systematic form but he also developed it to the extent of giving analytical solutions of linear and quadratic equations, which established him as the founder of Algebra.

In Arabic, however, algebra had long been applied to the solving of algebraic equations. The full Arabic expression was '*Ilm aljabr wa'l muqabalah*' (the science of reunion and equations), and the mathematician Al-Khawarizmi used *aljabr* as the title of his treatise on algebra. Al-Khwarizmi introduced the beginnings of algebra and wrote the first book of algebra, "*Hisab al-Jabr wal-Muqabalah*" (The Calculation of Integration and Equation). *Hisab al-Jabr wal-Muqabalah* contains analytical solutions of linear and quadratic equations, and its author may be called one of the founders of analysis or algebra as distinct from geometry. He also gives geometrical solutions (with figures) of quadratic equations, for example X2 + 1OX = 39, an equation often repeated by later writers.[19]

In the twelfth century, Gerard of Cremona and Roberts of Chester translated the algebra of Al-Khawarizmi into Latin. Mathematicians used it all over the world and it was taught in the principal textbooks of European universities until the sixteenth century. In it, he composes that given an equation, collecting the unknowns into one side of the equation is called Al-Jabr and collecting the known on the other side of the equation is called Al-Mukabalah. He also described many basic types of equations.[20] Algebra was a unifying theory that allowed rational numbers, irrational numbers, and geometrical magnitudes to all be treated as algebraic objects. It gave mathematics a whole new dimension and a development path, much broader in concept than before. It also enabled future development.

Another important aspect of the introduction of algebraic ideas was that it allowed mathematics to be applied to itself in a way that had not been possible earlier. Al-Khwarizmi also aided in announcing Arabic numerals, which are a decimal position system, and the concept of zero. Algebra and Algorithm are in fact combination of his work and name. Interestingly, this book on algebra comprised many examples from the Islamic inheritance laws and how they could be answered using algebra. Under Al-Mamun, the Caliph of that time, with some others, they were the first to map the globe.[21]

Several of his books were translated into Latin in the early twelfth century. In fact, his book on arithmetic, *Kitab Al-Jam'a wal Tafreeq bil Hisab Al-Hindi*, was lost in Arabic but survived in a Latin translation. Al-Khwarizmi was one of the first scholars to produce a detailed astronomical table (Zij), which was translated into European languages and, later, into Chinese. This astronomical table provided the means of calculating the positions of the stars and planets. His geography, captioned *Kitab Suraht-*

Al-Ard (The Face of the Earth), together with its maps, was also translated. This book inspired a generation of writers in Bagdad and Muslim Spain, or Al-Andalus, to unearth, analyze, and record geographical data. In addition, he wrote a book on the Jewish calendar, *Istikhraj Tarikh Al-Yahud*, and two books on the astrolabe. He also wrote *Kitab Al-Tarikh* and his book on sun-dials was captioned *Kitab Al-Rukhmat*, but both of them have been lost. He made many original contributions in geography, mathematics, and astronomy.[22]

His arithmetic synthesised Greek and Hindu knowledge and also contained his own contribution of fundamental importance to mathematics and science. Thus, he explained the use of zero, a numeral of fundamental importance developed by the Arabs. Similarly, he developed the decimal system so that the overall system of numerals, 'algorithm' or 'algorizm', is named after him. In addition to introducing the Indian system of numerals (now generally known as Arabic numerals), he developed at length several arithmetical procedures, including operations on fractions. It was through his work that the system of numerals was first introduced to the Arabs and later to Europe, through its translations in European languages. He also perfected the geometric representation of conic sections and developed the calculus of two errors, which practically led him to the concept of differentiation. He is also reported to have participated in the degree measurements ordered by Al-Mamun which were aimed at measuring the volume and circumference of the earth.[23]

In fact, the work of Al-Khwarizmi was a revolutionary move away from the Greek concept of mathematics, which was essentially based on geometry. Al-Khwarizmi's work, in Latin translation, carried the Arabic numerals along with the mathematics to Europe, through Spain. He developed sine, cosine, and trigonometry tables, which were later translated into the West.

Abu Yusuf Yaqub Ibn Ishaq al-Kindi (801-873)

Al-Kindi, a philosopher of Mutazilite theology, wielded considerable expertise in science and philosophy and was a descendant of the aristocratic family of Kinda. Al-Kindl was the first of the Muslim philosopher-scientists. He was the first great exponent of Greek philosophy in Islam and was deeply involved with the translation movement. His father, Ishak bin As-Sabbah, was the governor of Kufa under Al-Mahdi, Al-Hadi, and Al-Harun. Al-Kindi completed his early studies at Kufa, which was a centre for Arab culture and learning in the nineth century. His interests were encyclopedic. After that, Al-Kindi moved to Bagdad to complete his studies, and there he quickly achieved fame for his scholarship.

He came to the attention of the Caliph Al-Mamun, who was at that time setting up the "House of Wisdom" in Bagdad. Al-Kindi enjoyed the patronage of the Caliphs Al-Mamun, Al-Mu'tasim and Al-Mutawakkil and influenced them with the versatility of his genius and the profoundness of his knowledge. He wrote about two hundred seventy treatises, most of which are now lost, in logic, philosophy, physics, all branches of mathematics, as well as music, medicine, and natural history. He was the founder of the Islamic Peripatetic School of Philosophy. Cardano called him one of the twelve great intellectual figures of humanity.[24]

Al-Mamun was a patron of learning and founded an academy called the 'House of Wisdom' where Greek philosophical and scientific works were translated. Al-Kindi was appointed by Al-Mamun to the House of Wisdom together with Al-Khwarizmi and the Banu Musa brothers. The main task that Al-Kindi and his colleagues undertook in the 'House of Wisdom' the translation of Greek scientific manuscripts. Al-Mamun had built up a library of manuscripts, the first major library to be set up since that at Alexandria, collecting important works from Byzantium. In addition to the 'House of Wisdom', Al-Mamun set up observatories in which Muslim astronomers could build on the knowledge acquired by earlier peoples.[25]

He devoted himself mainly to translations, and is said to have translated numerous works on all sorts of subjects. A polymath philosopher, Al-Kindi, described frequency analysis and laid the foundation of cryptography. Al-Kindi remained in high positions throughout his life, except for the final years when court intrigues led to a misunderstanding between him and Caliph Mutawakkil. When he died, Al-Kindi's fame spread throughout the Muslim world, and he was honoured with the title of "the philosopher of the Arabs". After his death, the golden period of free inquiry ended, and the reactionary period of persecution began.

Al-Kindi, who is universally recognised as the independent thinker and first true Muslim philosopher, rejected the eternity of matter and the universe, despite the deep influence of Aristotle and Plotinus on his thought. He wrote on philosophy, mathematics, astronomy, medicine, politics, music etc. Versed in the languages of the Greeks, the Persians, and the Indians, and thoroughly acquainted with their sciences and philosophy, he was selected by Mamun for the work of translating Aristotle and other Greek writers into Arabic. He was a prolific writer; the total number of books written by him was 241, the prominent among which were divided as: Astronomy 16, Arithmetic 11, Geometry 32, Medicine 22, Physics 12, Philosophy 22, Logic 9, Psychology 5, and Music 7. It is unfortunate that most of his books are no longer extant, but those that do speak very highly of his standard of scholarship and contribution. He was known as Alkindus in Latin, and a large number of his books were translated into Latin by Gherard of Cremona. His books that were translated into Latin during the Middle Ages comprise *Risalah dar Tanjim*, *Ikhtiyarat al-Ayyam*, *Ilahyat-e-Aristu*, *Al-Mosiqa*, *Mad-o-Jazr*, and *Aduiyah Murakkaba*.[26]

He was the founder of the Islamic Peripatetic School of philosophy. As a rationalist and Mu'tazilite, Al-Kindi proposed that certain passages from the Holy Quran whose literal interpretation would be in conflict with reality should, instead, be understood as allegories to guide men of reason.[27] Al-Kindi believed that there existed two truths in the Holy Quran, one for the stupid and uneducated masses, and the other for the cultured and educated. Al-Kindi was of the opinion that the former were only able to appreciate simple things and so had to be enticed by the vision of hour is (related to Hoor) and other physical allurements. On the other hand, the latter were given the gifts of logic and reason so that they might arrive at a deeper meaning of the Holy Quran.[28]

In the court of Caliph Al-Mamun, Al-Kindi was a star who shone bright in the foremost cultural centre of the world. His academic pursuits maintained their vigour in the subsequent reigns of the rationalist Caliph Al-Mutassim and then of Al-Wathiq. But then came the ascendancy of the orthodox Caliph Al-Mutawwakil, who ended the long period of liberalism. Mutawwakil soon ordered the confiscation of Al-Kindi's personal library, known to all Bagdad as Al-Kindiyah.[29]

Thabit Ibn Qurra Ibn Marwan al-Sabi al-Harrani (836-901)

Thabit Ibn Qurra, born at Harran (present Turkey), belonging to Sebian sect of star worshipper, joined the scientific group at Bagdad that was being patronised by the Abbasid Caliphs. In Bagdad, Thabit received mathematical training and also training in medicine, which was common for scholars of that time. Thabit contributed to several branches of science, notably mathematics, astronomy, and mechanics, in addition to translating a large number of works from Greek to Arabic. Later, he was patronised by the Abbasid Caliph Al-Mutadid and was appointed court astronomer in Bagdad.[30]

Thabit, with his great skills in languages as well as great mathematical skills, translated and revised many of the important Greek works. He was instrumental in extending the concept of traditional geometry to geometrical algebra and proposed several theories that led to the development of non-Euclidean geometry, spherical trigonometry, integral calculus, and real numbers. Thabit was a brilliant scholar who made many important mathematical discoveries. He played an important role in preparing the way for such important mathematical discoveries as the extension of the concept of number to (positive) real numbers, integral calculus, theorems in spherical trigonometry, analytic geometry, and non-Euclidean geometry. In astronomy, Thabit was one of the first reformers of the Ptolemaic system, and in mechanics he was a founder of statics.[31]

Another important aspect of Thabit's work was his book on the composition of ratios. In this Thabit deals with arithmetical operations applied to ratios of geometrical quantities. Thabit also played a very important role in the establishment of astronomy as an exact science (method, topics, and programme), which developed along three lines: the theorisation of the relation between observation and theory; the 'mathematisation' of astronomy; and the focus on the conflicting relationship between 'mathematical' astronomy and 'physical' astronomy.[32]

In the field of mechanics, Thabit wrote an important work, *Kitab fi 'l-Qarastun* (The book on the beam balance). This book was translated into Latin by Gherard of Cremona and became a popular work on mechanics. In this work Thabit proves the principle of equilibrium of levers. He demonstrates that two equal loads, balancing a third, can be replaced by their sum placed at a point halfway between the two without destroying the equilibrium. After giving a generalization, Thabit then considers the case of equally distributed continuous loads and finds the conditions for the equilibrium of a heavy beam.[33]

Abu al-Hasan Ali bin Sahl Rabban al-Tabari (838-870)

Ali bin Rabban's father, Sahl, hailed from a respectable Jewish family. Ali hailed from an Israelite family, embraced Islam and classified amongst Muslim scholars. He was an accomplished Hakim and tutor of the unparalleled physician Zakariya al-Razi. Professionally, Sahl was an extremely successful physician. Ali received his education in the disciplines of medical science and calligraphy from his able father Sahl and attained perfection in these fields. He had also mastered the Syriac and Greek languages to a high degree of proficiency.[34]

The fame acquired by Ali Bin Rabban was due to his exaltation, which is found in his world-renowned treatise *Firdous al-Hikmat*. Spread over seven parts, *Firdous al-Hikmat* is the first ever medical encyclopaedia which incorporates all the branches of medical science into its folds. This work has been published in the twentieth century. Prior to this publication only five of his manuscripts were to be found scattered in libraries the world over. Dr. Mohammed Zubair Siddiqui compared and edited the manuscripts. In his preface he has provided extremely useful information regarding the book and the author and, wherever felt necessary, explanatory notes have been written to facilitate publication, of this work on modern publishing standards.[35]

Later on, this unique work was published with the cooperation of English and German institutions. The following are the details of its seven parts:[36]

1. Part one: Kulliyat-e-Tibb. This part throws light on the contemporary ideology of medical science. In that era, these principles formed the basis of medical science.
2. Part two: Elucidation of the organs of the human body, rules for keeping good health and comprehensive account of certain muscular diseases.
3. Part three: Description of the diet to be taken in conditions of health and disease.
4. Part four: All diseases right from head to toe. This part is of profound significance in the whole book and comprises twelve papers:
   (i) General causes relating to eruption of diseases.
   (ii) Diseases of the head and the brain.
   (iii) Diseases relating to the eye, nose, ear, mouth and the teeth.
   (iv) Muscular diseases (paralysis and spasm).
   (v) Diseases of the regions of the chest, throat and the lungs.
   (vi) Diseases of the abdomen.
   (vii) Diseases of the liver.
   (viii) Diseases of gallbladder and spleen.
   (ix) Intestinal diseases.
   (x) Different kinds of fever.
   (xi) Miscellaneous diseases—brief explanation of organs of the body.
   (xii) Examination of pulse and urine. This part is the largest in the book and is almost half the size of the whole book.
5. Part five: Description of flavour, taste and colour.
6. Part six: Drugs and poison.
7. Part seven: Deals with diverse topics, discusses climate and astronomy and also contains a brief mention of Indian medicine.

Abu al-Abbas Ahmad Ibn Muhammad Ibn Kathir al-Farghani (800-870)

Al-Farghani, born in Farghana, was one of the most distinguished astronomers in the service of al-Mamun and his successors. He was involved in the calculation of the diameter of the earth by the measurement of the meridian arch length, together with a team of scientists under the patronage of the Abbasid Caliph Al-Mamun in Bagdad. He determined the diameter of the earth to be 6,500 miles, and found the greatest distances and also the diameters of the planets.[37]

He wrote "Elements of Astronomy" (*Kitab fi al-Harakat al Samawiya wa Jawami Ilm al-Nujum,* i.e. the book on celestial motion and thorough science of the stars), "The Book of Chapters, a summary of the Almagest" (*Kitab al-Fusul, Ikhtiyar al-Majisti*) and "Book on the Construction of Sun-dials" (*Kitab Amal al-Rukhamat*). The *Jawami*, or 'The Elements' as we shall call it, was Al-Farghani's best-known and most influential work. It was translated into Latin in the 12th century and exerted a great influence on European astronomy before Regiomontanus. Jacob Anatoli produced a Hebrew translation of the book that served as a basis for a third Latin version, appearing in 1590, whereas Jacob Golius published a new Latin text together with the Arabic original in 1669. The influence of 'The Elements' on mediaeval Europe is clearly vindicated by the presence of innumerable Latin manuscripts in European libraries.[38]

Abu Bakr Mohammad Ibn Zakariya al-Razi (865-925)

Al-Razi, identified as Rhazes, was born in Ray, Iran. Initially, he was interested in music, but later on he learned medicine, mathematics, astronomy, chemistry, and philosophy. He became a student of Hunayn Ibn Ishaq and later a student of Ali Ibn Rabban. The practical experience gained at the well-known Muqtadari Hospital helped him in his chosen profession of medicine. At an early age, he gained eminence as an expert in medicine and alchemy, so that patients and students flocked to him from distant parts of Asia. Al-Razi was a Hakim, an alchemist and a philosopher. In medicine, his contribution was so significant that it can only be compared to that of Ibn Sina.

He was first placed in-charge of the first royal hospital at Ray, from where he soon moved to a similar position in Bagdad, where he remained the head of its famous Muqtadari Hospital for some time. He established separate wards in hospitals for the mentally ill, thereby creating the means for clinical observations of these diseases. He significantly preferred cure through accurate and controlled nourishment intake and studies the ideas involving human behaviour. This was shared with him, emphasising the impact of psychological aspects on health. In this way, he was a pioneer in the field of psychology, thus removing the theories of demons and witchcraft associated with these diseases in the Christian world.[39]

Some of his works in medicine, e.g. *Kitab al-Mansoori*, *Al-Hawi*, *Kitab al-Mulooki,* and *Kitab al-Judari wa al-Hasabah,* earned everlasting fame. *Kitab al-Mansoori*, which was translated into Latin in the fifteenth century, comprised ten volumes and dealt exhaustively with Greco-Arab medicine. Some of its volumes were published separately in Europe. His *Al-Judari wal Hasabah* was the first treatise on small pox and chicken pox, and is largely based on Razi's original contribution: It was translated into various European languages. Through this treatise, he became the first to draw clear comparisons between small pox and chicken pox. *Al-Hawi* was the largest medical encyclopaedia composed by then. It contained on each medical subject all the important information that was available from Greek and Arab sources, and this was concluded by him by giving his own remarks based on his experience and views.[40]

In addition to being a physician, he developed his chemistry independently of Jabir Ibn Hayyan. He has portrayed in great detail several chemical reactions and also given full descriptions of and designs for about twenty instruments used in chemical investigations. His description of chemical knowledge is in plain and plausible language. His books, called *Kitab-al-Asrar* (The Secrets) and *Sirr al-Asrar* (The Secret of Secrets), deals with the preparation of chemical materials and their utilization. Another one was translated into Latin under the name *Liber Experimentorum.* As a chemist, he was the first to produce sulphuric acid together with some other acids; and he also prepared alcohol by fermenting sweet products. Al-Razi set up a modern laboratory, designing and using more than 20 instruments like the crucible and still.[41]

He was a prolific author, who left behind monumental treatises on numerous subjects. He has more than 200 outstanding scientific contributions to his credit, out of which about half deal with medicine and another 21 with alchemy. He also wrote on physics, mathematics, astronomy, and optics, but these writings could not be preserved. A number of his books, including *Jami-fi-al-Tib*, *Mansoori*, *Al-Hawi*, *Kitab al-Jadari wa al-Hasabah*, *Al-Malooki*, *Maqalah fi al- Hasat fi Kuli wa al-Mathana*, *Kitab al-Qalb*, *Kitab al-Mafasil*, *Kitab-al-Ilaj al-Ghoraba*, *Bar al-Sa'ah*, and *Al-Taqseem wa al-Takhsir*, have been published in various European languages. About 40 of his manuscripts are still extant in the museums and libraries of Iran, Paris, Britain, Rampur, and Bankipur. His contributions have greatly influenced the development of science in general and medicine in particular.[42]

Al-Razi was more radical than Al-Kindi in his attachment to Greek rationalism. He openly spoke of the superiority of reason to revelation, for which he was charged with blasphemy and heresy. He received a severe punishment that he should be hit on the head with his books until either the book or his head broke. Thereupon Al-Razi lost both his eyesight, as well as his zest for life and shortly after died.[43]

Al-Razi, a Persian origin greatest clinical physician of Islam, earned the title of 'Arabic Galen' and 'most brilliant genius of the Middle Ages' for his phenomenal achievements in the Field of medicine. He created a treatment for kidney and bladder stones and clarified the nature of various infectious diseases. He also anticipated therapies first on animals in order to assess their effects and side effects. He was also an expert surgeon and the first to use opium for anesthesia.

Abu Abdallah Mohammad Ibn Jabir Ibn Sinan al-Raqqi al-Harrani al-Sabi al-Battani (858-929)

Al-Battani, born in Harren, was a Mesopotamian Arab astronomer, mathematician, and astrologer. Battani was first educated by his father, Jabir Ibn Sinan al-Battani, who was also a well-known scientist. He then moved to Raqqa, situated on the bank of the Euphrates, where he received advanced education and, later on, flourished as a scholar. At the beginning of the 9th century, he migrated to Samarra, where he worked till the end of his life. He was of Sabian origin (a star worshipper), but was himself a Muslim.

In the *Fihrist* or Index work, compiled by the bookseller Ibn an-Nadim in 988, he describes Al-Battani as follows:[44]

> "...one of the famous observers and a leader in geometry, theoretical and practical astronomy, and astrology. He composed a work on astronomy, with tables, containing his own observations of the sun and moon and a more accurate description of their motions then that given in Ptolemy's "Almagest". In it, moreover, he gives the motions of the five planets, with the improved observations he succeeded in making, as well as other necessary astronomical calculations. ...Nobody is known in Islam who reached similar perfection in observing the stars and scrutinising their motions..."

The *Fihrist* also quotes a number of works by Al-Battani. There is his *Kitab al-Zij,* which is his major work on astronomy with tables. The work contained 57 chapters. It begins with a description of the division of the celestial sphere into the signs of the Zodiac and into degrees. The necessary background mathematical tools are then introduced, such as the arithmetical operations on sexagesimal fractions and the trigonometric functions. Chapter 4 contains data from al-Battani's own observations. Chapters 5 to 26 discuss a large number of different astronomical problems, to some extent following material from the Almagest. The motions of the sun, moon, and five planets are discussed in chapters 27 to 31, where the theory given is that of Ptolemy, but for Al-Battani, the theory appears less important than the practical aspects. Al-Battani's 16 chapters explain how his tables are to be read. Chapters 49 to 55 cover problems in astrology, while Chapter 56 discusses the construction of a sun dial, and the final chapter discusses the construction of a number of astronomical instruments.[45]

Al-Battani catalogued 489 stars. He refined the existing values for the length of the solar year, which he gave as 365 days, 5 hours, 48 minutes, 24 seconds, and of the seasons. He calculated 54.5″ per year for the precession of the equinoxes and obtained the value of 23.35″ for the inclination of the ecliptic. Another of Al-Battani's accomplishments is that he concluded how an annular solar eclipse occurs. He did this by observing that the radius between the Earth and the Sun changes throughout the year. This led him to arrive at the conclusion that when the Sun is farthest from the Earth, an annular solar eclipse occurs. He was the first to make this observation and inference.[46]

His *Kitab al-Zij* was frequently quoted by many mediaeval astronomers, including Copernicus. Often called the Ptolemy of the Arabs, Al-Battani is perhaps the greatest and best known astronomer of the mediaeval Islamic world. In mathematics, Al-Battani was the first to substitute the practice of Greek chords and the first to cultivate the concept of cotangent, and they provided their table in degrees. Al-Battani systematically developed trigonometry and extended it to spherical trigonometry, with significant implications for astronomy, geography, and exploration beyond the known world, allowing for the creation of better maps and the reconceptualization of the planet's structure.

Abu Nasr Mohammad Ibn al-Farakh al-Farabi (872-950)

Al-Farabi, known in the West as Alpharabius, so called from his native city of Farab in Turkistan, was a distinguished physician, mathematician, and philosopher. He completed his earlier education at Farab and Bukhara, but later moved to Bagdad for higher studies, where he studied and worked for a long time. He lived through the reigns of six Abbasid Caliphs. Farabi travelled to many distant lands and studied for some time in Damascus and Egypt, but repeatedly came back to Bagdad, until he visited Saif ud-Dowla Ali bin Hamdan's court in Halab (Allepo). He became one of the constant companions of the King, and it was here at Halab that his fame spread far and wide, and died at Damascus.

Farabi contributed considerably to science, philosophy, logic, sociology, medicine, mathematics, and music. As a philosopher, he may be classed as a Neoplatonist who tried to synthesise Platonism and Aristotelism with theology, and he wrote such rich commentaries on Aristotle's physics, meteorology, logic etc., in addition to a large number of books on several other subjects embodying his original contribution, that he came to be known as the "Second Teacher" (*Al-Mou'allim al-Thani*), Aristotle being the First. One of the important contributions of Farabi was to make the study of logic easier by dividing it into two categories, viz. *Takhayyul* (idea) and *Thubut* (proof).[47]

Although many of his books have been lost, 117 are known, out of which 43 are on logic, 11 on metaphysics, 7 on ethics, 7 on political science, 17 on music, medicine, and sociology, while 11 are commentaries. Some of his more famous books include the book *Fusus al-Hikam*, which remained a text book of philosophy for several centuries at various centres of learning and is still taught at some of the institutions in the East. The book *Kitab al-lhsa al Ulum* (Encyclopaedia of Science) discusses classification and fundamental principles of science in a unique and useful manner. The book *Ara Ahl al-Madina al-Fadila* (The Model City) is a significant early contribution to sociology and political science.[48]

His works deal with the different branches of science, viz. language, logic, mathematics, natural sciences, medicine, political and social economy, and philosophy. Besides philosophy and medicine, Farabi cultivated music, which he elevated into a science. He wrote several treatises both on the theory and the art of music, as well as on the manufacture of musical instruments. He compared the systems of music among the ancients with those in vogue in his own time.

Abul Hasan Ali Ibn Husain Ibn Ali ́al-Masudi (896-956)

Al-Masudi was a great Mutazilite Arab historian, physicist, geographer, and traveller, and was referred to as the "Herodotus and Pliny of the Arabs". By presenting a critical account of historical events, he initiated a change in the art of historical writing, introducing the elements of analysis, reflection, and criticism, which were later improved by Ibn Khaldun. In particular, in *Al-Tanbeeh* he makes a systematic study of history from the perspective of geography, sociology, anthropology, and ecology. Masudi had a deep insight into the causes of the rise and fall of nations.

He travelled to Fars and preceded India *via* Bagdad, where he visited Multan, Mansoora, and Kirman. In those days, Mansoora was a city of great renown and was the capital of the Muslim state of Sind. Masudi also travelled to Gujrat, and also travelled to the Deccan, Ceylon, Indo-China, and China, and proceeded *via* Madagascar, Zanjibar, and Oman to Basra. At Basra he completed his book *Muruj al-Dhahab wa-Maadin al-Jawhar* (The Meadows of Gold and Mines of Gems), in which he combines universal history with scientific geography, social commentary, and biography. In this book, Masudi described in a most absorbing manner his experience of various countries, peoples, and climates. In this book, he provides important information on early Arab music as well as music of other countries. He also gives accounts of his personal contacts with the Jews, Iranians, Indians, and Christians.[49]

From Basra, he moved to Syria and from there to Cairo, where he wrote his second extensive book, *Muruj al-Zaman* (Histotry of the Time) in thirty volumes. In this book he has described in detail the geography and history of the countries that he has visited. He also prepared a supplement, called *Kitab al-Ausat*, in which he compiled historical events chronologically. In the year of his death, he completed his last book, *Kitab al-Tanbih wa al-Ishraf*, in which he gave a summary of his earlier book as well as an errata.[50]

Whatever he has recorded was with a scientific approach and constituted an important contribution to geography, history, and the earth sciences. It is interesting to note that he was one of the early scientists who propounded several aspects of evolution, viz. from minerals to plants, plants to animals and animals to man. His research and views extensively influenced the sciences of historiography, geography, and earth sciences in several countries. Al-Masudi was one of the greatest geographers in the world, and his comprehensive account of lands of the Caliphate is invaluable to scholars. The knowledge of geography which these travels yielded is of the highest importance to science.

Abul Wafa Muhammad Ibn Muhammad Ibn Yahya Ibn Ismail al-Buzjani (940-998)

Abu al-Wafa Muhammad Al-Buzjani was born in Buzjan, Nishapur. He became a great mathematician, astronomer, and geometrician in Bagdad. Al-Buzjani's main contributions lie in several divisions of mathematics, specifically geometry and trigonometry.

In geometry, he added to the solution of geometrical problems with the opening of the compass, the construction of a square equivalent to other squares, regular polyhedral, the construction of a regular hexagon taking for its side the equilateral triangle inscribed in the same circle, the construction of parabola by points and the geometrical solution of the equations x4 = a and x4 + ax3 = b. Abul Wafa's contribution to the development of trigonometry was extensive. He was the first to show the generality of the sine theorem relative to spherical triangles. He developed a new method of constructing sine tables, the value of sin 30° being correct to the eighth decimal place. He also developed relations for sine (a+b) and the formula: 2 sin2 (a/2) = 1 – cos a, and sin a = 2 sin (a/2) cos (a/2). In addition, he made a special study of the tangent and calculated a table of tangents. He introduced the secant and cosecant for the first time, knew the relations between the trigonometric lines, which are now used to define them, and undertook extensive studies on conics.[51]

He wrote a large number of books on mathematics and other subjects, most of which have been lost or exist in modified forms. His contributions include *Kitab Ilm al-Hisab*, a practical book of arithmetic, *Al-Kitab al-Kamil* (the Complete Book), and *Kitab al-Handsa* (Applied Geometry). Apart from this, he wrote rich commentaries on Euclid, Diophantos, and al-Khawarizmi, but all of these have been lost.[52]

Abul Qasim Khalaf Ibn al-Abbas al-Zahrawi (936-1013)

Al-Zahrawi was born in the city of Azahara in the northwest of Cordoba, Andalusia. In European languages, his name is written in over a dozen different ways: Abulcasis, Albucasis, Bulcasis, Bulcasim, Bulcari, Alzahawi, Ezzahrawi, Zahravius, Alcarani, Alsarani, Aicaravi, Alcaravius, Alsahrawi etc. He was the greatest surgeon of the Middle Ages, known as the father of modern surgery. He became one of the most renowned surgeons of the Muslim era and was physician to King Al-Hakam-II of Spain. Al-Zahrawi's pioneering contributions to the field of surgical procedures and instruments had an enormous impact in the East and West well into the modern period, where some of his discoveries are still applied in medicine to this day.

Al-Zahrawi also pioneered neurosurgery and neurological diagnosis. He is known to have performed surgical treatments for head injuries, skull fractures, spinal injuries, hydrocephalus, subdural effusions, and headaches. The first clinical description of an operative procedure for hydrocephalus was given by Al-Zahrawi, who clearly describes the evacuation of superficial intracranial fluid in hydrocephalic children. Al-Zahrawi's thirty-volume medical encyclopaedia, *Kitab al-Tasrif* (Book of Concessions), completed in the year 1000, covered a broad range of medical topics, including surgery, medicine, orthopaedics, ophthalmology, pharmacology, nutrition, dentistry, childbirth, and pathology.[53] This book contained data that Al-Zahrawi had accumulated during a career that spanned almost 50 years of training, teaching, and practice. He apparently travelled very little but had wide experience in treating accident victims and war casualties.

Two of Al-Zahrawi's treatises deserve special mention. Firstly, his 28th treatise, known in Latin as *Liber servitoris de preparatione medicinarum simplicium*, describes chemical preparations, tablet making, filtering of extracts and related pharmaceutical techniques. This treatise was printed in Venice in 1471 by Nicolaus Jensen. Another, and perhaps the most important treatise, is the one on surgery. This monumental work was the first in Arabic to treat surgery independently and in detail. It included many pictures of surgical instruments, most invented by Al-Zahrawi himself, and explanations of their use. Al-Zahrawi was the first medical author to provide illustrations of instruments used in surgery. There are approximately 200 such drawings, ranging from a tongue depressor and a tooth extractor to a catheter and an elaborate obstetric device.[54]

The variety of operations covered is amazing. Al-Zahrawi discussed cauterisation, bloodletting, midwifery, obstetrics, and the treatment of wounds. He invented more than 200 surgical tools and devices for purposes such as inspection of the interior of the urethra and also for inspection, applying and removing foreign bodies from the throat, the ear and other body organs. These surgical tools and devices revolutionised medical science. He was also the first to illustrate the various cannulae and the first to treat a wart with an iron tube and caustic metal as a boring instrument.

Abu Ali Hasan Ibn al-Haitham (965-1041)

Al-Haitham was one of the most eminent physicists, great mathematicians and astronomers. Al-Haitham, known in the West as Alhazen or Alhacen, was born and educated in Basrah. He received, by all accounts, the best education that the Islamic world had to offer, studying in both Basra and Bagdad, which was arguably the greatest city in the world at the time. He seems to have excelled in mathematics and science and continued his studies in these fields even when he was working in a civil service position in Iran as a young adult.

Although he was famed during his lifetime for accomplishments and learning in a variety of subjects, what he is most known for today, and the area in which he had the greatest influence on the mediaeval world, was his theory of optics. His contributions to optics and scientific methods are outstanding. Al-Haitham spent the majority of his life in Cairo, the Fatimid capital. He made a thorough examination of the passage of light through various media and discovered the laws of refraction.

He solved the issue of how do we see the things that are around us? He proved that if things are visible, it's because they have sent some light that our eyes have received. The eye is, therefore, passive and sensitive to an outside agent, which he, to a certain degree, can control. Ibn al-Haitham held the anatomical descriptions of the eye to which he dedicated a full chapter in his optical treatise. His work laid the foundation for such later scholars as Johannes Kepler and Roger Bacon. Both of these scholars, and many others, were intrigued by Al-Haitham's experiments with mirrors. He also carried out the first experiments on the dispersion of light into its constituent colours.

He is more famous for his discovery of Camera (*Qamara* in Arabic—A darkened room). Al-Haitham observed light coming through a small hole in the window, shutters producing an upside-down image on the opposite wall. Al-Haitham said that vision was caused by light reflecting off an object and entering the eye. He wrote a seven-volume treatise on the subject, the *Kitab al-Manazir,* or Book of Optics, which was translated into Latin at the end of the 12th century and, thus, passed into the scholarly world of mediaeval Europe. His work "*Kitab al-Manazir*" made Europeans aware of Al-Haitham's amazing successes in the field of optics. He dealt at length with the theory of various physical phenomena like shadows, eclipses, and the rainbow and he speculated on the physical nature of light. He is the first to accurately describe the various parts of the eye and give a scientific explanation of the process of vision. He also attempted to explain binocular vision and he gave a correct explanation of the apparent increase in size of the sun and the moon when near the horizon. Through this extensive research on optics, he has been considered as the father of modern Optics.[55]

Al-Haitham proved that the refraction of a ray of light increases with the increasing density of the atmosphere; he also determined approximately the height of the layer of air surrounding the Earth. He brilliantly investigated the phenomenon of atmospheric refraction, calculating the height of the atmosphere to be 16 kilometres. This compares well with modern measurements of the troposphere, the lowest layer of the atmosphere, which measure 11 kilometres. He used experimental evidence to check his theories. He was the first to introduce experimental evidence as a requirement for accepting a theory. Al-Haytham realised that the senses were prone to error, and he devised scientific methods of verification, testing, and experimentation to uncover the truth about the natural phenomena he saw.[56]

Ibn al-Haitham discussed the density of the atmosphere and developed a relationship between it and height in his book *Mizan al-Hikmah*. He also studied atmospheric refraction. He discovered that the twilight only ceases or begins when the sun is 19° below the horizon and attempted to measure the height of the atmosphere on that basis. He has also discussed the theories of attraction between masses, and it seems that he was aware of the magnitude of acceleration due to gravity.[57]

In his book called, *The Balance of Wisdom,* he discusses dynamical principles, generally supposed to be the monopoly of modern science. He describes minutely the connection between the weight of the atmosphere and its density and how material objects vary in weight in a rare and dense atmosphere. He discusses the submergence of floating bodies and the force with which they rise to the surface when immersed in light or heavy media; he fully understands the principle of gravitation and recognises gravity as a force. He knows the correctly relations between the velocities, spaces, and times of falling bodies and has very distinct ideas of capillary attraction.[58]

Al-Haitham was summoned to Cairo by the ruling Fatimid Caliph Al-Hakim to see if he could help to control the Nile's unpredictable floods. Confident of his own abilities, Al-Haitham boasted that he would tame the great river by building a dam and reservoir. But, when he saw the extent of the challenge he had taken on, he realised it was impossible. To avoid the Caliph's wrath at his failure, Al-Haitham decided to pretend to be mad, and so, for his own protection, he was kept under effective house arrest in the Al-Azhar in Cairo. It was while imprisoned that Al-Haitham made the discovery of the camera for which he is best remembered.[59]

Al-Haitham also studied lenses, experimenting with different mirrors: flat, spherical, parabolic, cylindrical, concave, and convex. Al-Haytham was correct in his theories of vision and refraction of light, the position of the image produced by curved mirrors, and the seeming dimensions of objects. He completely changed the understanding of light and vision, although he paid a high price for his scientific honesty. Al-Haitham was the first to test his theories by means of rigorously controlled physical experiments.

Al-Haitham corrected the Greek misconception as to the nature of vision and demonstrated for the first time that the rays of light come from external objects to the eye, and do not issue forth from the eye and impinge on external things. He determined the retina as the seat of vision and proved that the impressions made upon it were conveyed along the optic nerves to the brain. He explained the phenomena of single vision by the formation of visual images on symmetrical portions of the two retinas. He discovered that the refraction of light varies with the density of the atmosphere, and that atmospheric density again varies with height. He explained accurately and clearly how, as a consequence of this refraction, astral bodies are seen before they have actually risen and after they have set, and he demonstrated that the beautiful phenomenon of twilight was due to the effect of atmospheric refraction combined with the reflecting action of the air upon the course of the rays of light.[60]

The list of his books runs to 200 or so, very few of which have survived. Even his monumental treatise on optics survived through its Latin translation. During the Middle Ages, his books on cosmology were translated into Latin, Hebrew, and other languages. He has also written on the subject of evolution, a book that deserves serious attention even today. Al-Haitham's discoveries were certainly important to the mediaeval world, but by far, his greatest achievement lies not in what he discovered but in how he set about discovering it. He has been called, of course, the father of modem optics, as well as the second Ptolemy, but perhaps his most impressive title is the father of scientific methodology.

Abu Ali al-Hussain Ibn Abdallah Ibn Sina (980-1037)

Ibn Sina, known as Avicenna in the West, born in Uzbekistan (Afshana near Bukhara), was a Persian Polymath and regarded as one of the most significant physicians and the father of early modern medicine. By age 10, he had memorised all of the Quran. At 13, he started to study medicine in earnest, and after that, he studied law. At age 16, he embarked on the study of philosophy. Ibn Sina taught widely in medicine, philosophy, and the natural sciences. Ibn Sina was honorifically called Al-Shaykh al-Ra'is, the Grand Shaikh, Prince of Physicians. Ibn Sina gave a systematic account of medicine, and his work on medicine remained unchallenged for centuries. Ibn Sina devoted his life to the study of medicine, philosophy, and other branches of science.

While still young, he attained such a degree of expertise in medicine that his renown spread far and wide. Even as a young man, he was regularly consulted by high-ranking authorities on matters of medicine, religion, and philosophy. For example, at the age of 17, he was fortunate in curing Samani ruler Nuh Ibn Mansoor, the King of Bukhhara, of an illness in which all the well-known physicians had given up hope. The youth promptly cured the emir, who then, in gratitude, offered Avicenna the use of his vast library. Ibn Sina went to Hamadan, where he cured Amir Shamsud-Dawala of colic and was made Prime Minister.

Al-Qifti states that Ibn Sina completed 21 major and 24 minor works on philosophy, medicine, theology, geometry, astronomy and the like. Another source (Brockelmann) attributes 99 books to Ibn Sina, comprising 16 on medicine, 68 on theology and metaphysics, 11 on astronomy, and four on verse. Another source mentions that Ibn Sina wrote some 246 books dealing with philosophy, medicine, geometry, astronomy, theology, philosophy, and art. Most of these were in Arabic; but in his native Persian he wrote a large manual on philosophical science entitled *Danish-naama-i-Alai* and a small treatise on the pulse. His most celebrated Arabic poem describes the descent of the soul into the body from the higher sphere. Among his scientific works, the leading two are the *Kitab al-Shifa* (Book of Healing), a philosophical encyclopaedia based upon Aristotelian traditions, and the *Al-Qanun fi al-Tibb* (The Canon of Medicine), which represents the final categorisation of Graeco-Arabian thoughts on Medicine.[61]

Apart from his renowned book, *Al-Qanun fi al-Tibb,* the other books, *Kitab al-Shifa*, *Al-Najat* (The Book of Salvation), *Al-Mahda wa'l Maad* (The Beginning and the End), and the *Isharat* (The Book of Directives and Remarks), are also considered the most important and principal guides in Asia and Europe from the twelfth to seventeenth centuries. In his *Kitab al-Shifa,* containing 20 volumes, he divided practical knowledge into ethics, economics, and politics, and theoretical knowledge into mathematics, physics, and metaphysics.[62]

The Qanun, or Canon of Medicine, is, of course, by far the largest, most famous, and most important of Ibn Sina's works. The work contains about one million words and, like most Arabic books, is elaborately divided and subdivided. The main division is into five books, of which the first deals with general principles; the second with simple drugs arranged alphabetically; the third with diseases of particular organs and members of the body from the head to the foot; the fourth with diseases, which though local in their inception, spread to other parts of the body, such as fevers; and the fifth with compound medicines. He also provided a list of how to prepare, dispense, and properly use more than 800 medical compounds to treat various ailments.[63]

The Canon of Medicine distinguishes mediastinitis from pleurisy and recognises the contagious nature of phthisis (tuberculosis of the lung) and the spread of disease by water and soil. It gives a scientific diagnosis of ankylostomiasis and attributes the condition to an intestinal worm. The Qanun points out the importance of dietetics, the influence of climate and the environment on health, and the surgical use of oral anaesthetics. Ibn Sina advised surgeons to treat cancer in its earliest stages, ensuring the removal of all the diseased tissue. Avicenna offered advice on how to use anaesthetics when performing surgery. The Qanun's Materia Medica considers some 760 drugs, with comments on their application and effectiveness. He recommended the testing of a new drug on animals and humans prior to general use.[64]

Ibn Sina noted the close relationship between emotions and the physical condition and felt that music had a definite physical and psychological effect on patients. Of the many psychological disorders that he described in the Qanun, one is of unusual interest: love sickness! Ibn Sina is said to have identified this condition in a prince in Jurjan who was sick and whose illness had perplexed local doctors. Ibn Sina noted a fluttering in the Prince's pulse when the address and name of his beloved were mentioned. The great doctor had a simple remedy: unite the sufferer with the beloved.[65]

The Canon of Medicine was translated into Latin in the twelfth century, and it was used in medical schools throughout Europe until the advent of modern science. The Canon of Medicine contained all Greek medical knowledge together with Arabic interpretations and contributions. Medical progress did not shake the Canon of Medicine, and there was still a professional market for the Canon during the whole of the seventeenth century. He also developed methods of treating fractured bones that doctors still adhere to today. Medical books written by these men and other scholars in Muslim civilization influenced European medicine for centuries.

He established free hospitals and developed treatments for diseases using herbs, hot baths, and even major surgery. Due to their medical and medicinal work, the Ottomans were particularly noted for their building of hospitals and for the high level of hygiene practiced in them. Every city in the Islamic world had a number of excellent hospitals, many of which were specialised for specific diseases, including mental and emotional disorders. Ibn Sina says:[66]

> "If a problem was too great for me, I repaired to the mosque and prayed, invoking the Creator of all things, until the gate that had been closed to me was opened and what had been complex became simple. Always, as night fell, I returned to my house, set the lamp before me, and buried myself in reading and writing. If sleep overcame me or I felt the flesh growing weak, I had recourse to a beaker of wine, so that my energies were restored."

Ibn Sina was a fiercely independent-minded philosopher who insisted on the primacy of reason. For a time, he was a minister to the Amir of Hamedan. Here he got into a religious argument with strict believers in the army, and they soon called for his execution. Soldiers came to his house but, not finding him, plundered it and then called on the Amir to behead him. Ibn Sina was warned in time and hid at the house of his friend Abu

Said Dafdaq, where he worked on his masterpiece, *Al-Qanun*. He fled persecution and the wrath of rulers several times. His book was banned and he was branded as heresy.[67]

Ibn Sina is considered by many to be the greatest thinker and polymath of the Islamic golden age, and his writings were deeply influential on the great intellects of the Islamic, Christian, and Jewish traditions. Arguably, his greatest achievement was to fuse Islamic theology with Aristotelian metaphysical thought, although in the West, it was his encyclopaedic medical texts that had the most profound and visible impact on mediaeval society. Ibn Sina is still known in the world and taught various courses of medicine. In the museum at Bukhara, there are displays showing many of his writings, surgical instruments from the period; and paintings of patients undergoing treatment. An impressive monument to the life and works of the man who became known as the 'doctor of doctors' still stands outside the Bukhara museum, and his portrait hangs in the Hall of the Faculty of Medicine at the University of Paris.

Abu Raihan Mohammad Ibn Ahmad al-Biruni (973-1049)

Al-Biruni was born in the outer district (Birun) of Khiva, near Oxus or Amu Darya, in Khwarizm (now in Uzbekistan). He was one of the well-known figures associated with the court of Khwarizm, ruler of the Afrighid dynasty and Ghazni afterwards under Mahmud Ghaznavi. Sultan Mahmud Ghaznavi took al-Biruni along with him on his journeys to India several times, and thus he had the opportunity to travel all over India during a period of 20 years. He learnt Hindu philosophy, mathematics, geography and religion from the Pandits to whom he taught Greek and Arabic science and philosophy.

Al-Biruni was a versatile scholar and scientist who had equal facility as a physicist, pharmacologist, mathematician, astronomer, astrologer, natural scientist, historian, chronologist, and linguist. He was conversant in Khwarezmian, Persian, Arabic, and Sanskrit, and also knew Greek, Hebrew, and Syriac. He has been variously called the "founder of Indology", "Father of Comparative Religion", the "Father of modern geodesy", and the first anthropologist. He was honoured by his contemporaries by the honorific title of "the Master" (Al-Ustadh), but was unknown to the mediaeval West, except perhaps by the garbled name Maitre Aliboron.

He recorded observations of his travels through India in his well-known book *Kitab al-Hind,* which gives a graphic account of the historical and social conditions of the subcontinent. At the end of this book he makes a mention of having translated two Sanskrit books into Arabic, one called *Sakaya*, which deals with the creation of things and their types, and the second, *Patanjal,* dealing with what happens after the spirit leaves the body. His descriptions of India were so complete that even the *Aain-i-Akbari,* written by Abul Fazal during the reign of Akbar, 600 years later, owes a great deal to Al-Biruni's book. He observed that the Indus valley must be considered as an ancient sea basin filled up with alluvials.[68]

On his return from India, Al-Biruni wrote his famous book *Qanun-i Masoodi* (*Al-Qanun al-Masudi, fi al-Hai'a wa al-Nujum*), which he dedicated to Sultan Masood. The book discusses several theories of astronomy, trigonometry, solar, lunar, and planetary motions, and related topics. In another well-known book, *Al-Athar al-Baqia*, he has attempted a connected account of the ancient history of nations and the related geographical knowledge. In this book, he has discussed the rotation of the earth and given correct values of the latitudes and longitudes of various places. He has also made a considerable contribution to several aspects of physical and economic geography in this book.[69]

His other scientific contributions include the accurate determination of the densities of 18 types of precious stones. His book, the *Kitab al-Jamahir fi-marifat al-Jawahir* (The Book Most Comprehensive in Knowledge on Precious Stones), deals with the properties of various precious stones. Al-Biruni describes various precious stones, narrates events from his travels, quotes poets, story-tellers, and philosophers, all the while describing the value or properties of precious stones and gems. He also gives scientific descriptions of the way certain precious stones are formed.[70]

He also wrote the *Kitab-al-Saidana*, which is an extensive materia medica that combines the then-existing Arabic knowledge on the subject with Indian medicine. He was also an astrologer and is reputed to have astonished people by the accuracy of his predictions. He gave a clear account of Hindu numerals, elaborating on the principle of position.[71]

His most important works are the inclusion of Arabic nomenclature of shade and shadows; strange phenomena involving shadows; gnomonic; the history of the tangent and secant functions; applications of the shadow functions to the astrolabe and to other instruments; shadow observations for the solution of various astronomical problems; and the shadow-determined times of Muslim prayers. Al-Biruni established the rule which stated that the specific density of a body is equivalent to the volume of the water which makes it move. He also interpreted the exit of water from geysers and artesian wells in light of the theory of communicating vessels. Al-Biruni was more successful, and special mention should be made here of Al-Biruni's theory of specific gravity. He determined the specific gravity of nine metals and nine precious stones. He defined the ratios between the densities of gold, mercury, lead, silver, bronze, copper, brass, iron, and tin.[72]

It is believed that Al-Biruni wrote 146 books and treatises. Apart from *Kitab al-Hind* (History and Geography of India), *Al-Qanun al-Masudi* (Astronomy, Trigonometry), *Al-Athar al-Baqia* (Ancient History and Geography), *Kitab al-Saidana* (Materia Medica), and *Kitab al-Jawahir* (Precious Stones), his book *Al-Tafhim-li-Awail Sina'at al-Tanjim* gives a summary of mathematics and astronomy. In his *Kitab Tahdid al-Amakin,* Al-Biruni described a simple method for architects and artisans to determine *Qiblah* from Ghazna. In his book *Tahdid,* Al-Biruni describes has observational abilities and formulation of geological theories:[73]

> "We do not know of the conditions of creation, except what is observed in its colossal and minute monuments, which were formed over long periods of time. Foe example, the high mountains, which are composed of soft fragments of rocks of different colours, combined with clay and sand, which have solidified over their surfaces. A thoughtful study of this matter will reveal that the fragments and pebbles are stones which were torn from the mountains by internal splitting and by external collision. The stones then wear off by the continuous friction of enormous quantities of water that run over them and by the wind that blows over them. This wearing off takes place first at the corners and edges, until they are rubbed off and the stones finally take an approximate spherical shape. As a contradistinction to the mountains, we have the minute particles of sand and earth."

Al-Biruni had a very independent mind and would not accept anything merely on authority. Although he lived at a time when the impact of the translation movement was the strongest, he rejected the arguments of the Greek philosophers on the eternity of the world. His cosmological doctrines reveal an amazing synthesis of his observations during travel, his readings, and scientific experiments. Al-Biruni gives geographical details of the various regions of the Earth; describes the formation of mountains; clouds; dry and wet lands; the seven climes; and numerous minerals, plants, and animals. In his writings on the animal and plant kingdoms, his ideas are directly inspired by the Quran, just as his overall scheme of human existence and the human relationship with the world.[74]

He has been considered as one of the very greatest scientists of Islam and, all considered, one of the greatest of all time. His critical spirit, love of truth, and scientific approach were combined with a sense of toleration. His enthusiasm for knowledge may be judged from his claim that the phrase Allah is Omniscient does not justify ignorance.

Ghiyath al-Din Abul Fateh Umar Ibn Ibrahim al-Khayyam (1048-1131)

Umar Khayyam, born in Nishapur, was a Persian polymath, mathematician, astronomer, philosopher, historian, physician, and poet. It has also been said that he could have belonged to the Khayyami tribe of Arab origin who might have settled in Persia. He travelled to the great centres of learning, Samarqand, Bukhara, Balkh, and Isphahan, in order to study further and exchange views with the scholars there.

It was a very difficult time for the men of learning under the Seljuks. Khayyam himself described the difficulties for men of learning during this period in the introduction to his "Treatise on Demonstration of Problems of Algebra" (*Maqalat fi al-Jabr wa al-Muqabila*). In metaphysics, he wrote three books: *Risala Dar Wujud,* and the recently discovered *Nauruz-namah*. He was also a renowned astronomer and a physician.[75]

No account of mathematics in Islamic civilization can be complete without the mention of Umar Khayyam, who is most known in the West for his *Rubaiyat* (quatrains). Khayyam wrote an undiscovered treatise called Problems of Arithmetic (*Mushkalat al-Hisab*), a key treatise on cubic equations (the *Risala*), a lengthy commentary on Euclid, and many other works on astronomy, music, arithmetic, and algebra, in addition to his better known poetic and philosophical works.[76]

The Seljuk ruler Malik-Shah sent an invitation through his vizier Nizam al-Mulk asking Khayyam to go to Esfahan to set up an observatory. Other leading astronomers were also brought to the Observatory in Esfahan, and for 18 years, Khayyam led the scientists and produced work of outstanding quality. He became the head of the team of the most distinguished astronomers of the eleventh century. It was a period of peace during which the political situation allowed Khayyam the opportunity to devote himself entirely to his scholarly work. He compiled *Zij Malik-Shahi* (Malik-Shah Astronomical Tables) at the observatory in Isfahan.[77]

Khayyam introduced a calendar that was remarkably accurate and was named *Al-Tarikh-al-Jalali*. It had an error of one day in 3770 years and was thus even superior to the Georgian calendar (an error of one day in 3330 years). Khayyam measured the length of the year as

365.24219858156 days. There may be two comments on this result. Firstly, it shows incredible confidence to attempt to give the result to this degree of accuracy. We know now that the length of the year is changing to the sixth decimal place over a person's lifetime. Secondly, it is outstandingly accurate. For comparison, the length of the year at the end of the 19th century was 365.242196 days, while today it is 365.242190 days.[78]

Outside the world of mathematics, Khayyam is best known as a result of Edward Fitzgerald's popular translation in 1859 of nearly 600 short four-line poems, the Rubaiyat. Khayyam's fame as a poet has caused some to forget his scientific achievements which were much more substantial. Versions of the forms and verses used in the Rubaiyat existed in Persian literature before Khayyam, and only about 120 of the verses can be attributed to him with certainty. Of all the verses, the best known is the following:[79]

> The Moving Finger writes, and, having writ,
> Moves on: nor all thy Piety nor Wit
> Shall lure it back to cancel half a Line,
> Nor all thy Tears wash out a Word of it.

Abu Bakr Mohammad Ibn Yahya Ibn ul-Sayegh (1095-1138)

Ibn Yahya Ibn ul-Sayegh, popularly called Ibn Bajjah, is one of the most celebrated philosophers among the Arabs of Spain. He was not only a distinguished physician, mathematician, and astronomer, but also a musician of the first rank. He was born in Saragossa. We find him mentioned as residing in Seville. He afterwards proceeded to Africa, where he occupied a high position under the Almoravids and died at Fez. Several of his works have been preserved in their entirety, demonstrating the freedom of the Muslim intellect in those days.[80]

Abu Bakr Mohammad Ibn Abdul Malik Ibn Tufail al-Kaisi (1105-1185)

Ibn Tufail (Abubacer) was born in Wadi Ash, a small town northeast of Granada in Andalusia. He was celebrated as a physician, mathematician, astronomer, philosopher, and poet, and was held in great esteem at the court of the first two sovereigns of the Almohade dynasty. From 1163 to 1184, he filled the office of Vazir and physician to Abu Ya'kub Yusuf, the second Almohade king. His most famous work was his philosophical novel, *Hayy Ibn Yakzan* (Living the Son of the Awake), which represents the gradual and successive development of intelligence and the power of perception in a person wholly unassisted by outside instruction. As a physician, he was an early supporter of dissection and autopsy, which were expressed in his novel.[81]

Ibn Tufail's imaginative tale is an attempt to show that knowledge of the truth gained through the intellect and reason does not contradict what is apprehended through mystical experience. The tale is based on an earlier work of the same title by Ibn Sina. It is rooted in the Islamic concept of *Fitrah*, the innate nature of human beings, and in the Quranic message that when left to itself, the human intellect is capable of comprehending the ultimate reality, which is none other than One Allah.

Abu Marwan Abd al-Malik Ibn Zuhr (1091-1161)

Ibn Zuhr was born in Seville and completed his education and specialisation in medicine. Ibn Zuhr was one of the greatest physicians and clinicians of the Muslim golden era and has rather been held by some historians of science as the greatest of them. Contrary to the general practice of the Muslim scholars of that era, he confined his work to only one field of medicine. This enabled him to produce works of everlasting fame.

As a physician, he made several discoveries and breakthroughs. He described correctly, for the first time, scabies, the itch mite, and may thus be regarded as the first parasitologist. Likewise, he prescribed tracheotomy and direct feeding through the gullet and rectum in cases where normal feeding was not possible. He also gave clinical descriptions of mediastinal tumours, intestinal phthisis, inflammation of the middle ear, pericarditis etc.[82]

His contribution was chiefly contained in the monumental works written by him; of these, however, only three are extant. *Kitab al-Taisir fi al-Mudawat wa al-Tadbir* (Book of Simplification concerning Therapeutics and Diet), written at the request of Ibn Rushd (Averroes), is the most important work of Ibn Zuhr. It describes several original contributions of Ibn Zuhr. The book gives in detail the pathological conditions, followed by therapy. His *Kitab al-Iqtisad fi Islah al-Anfus wa al-Ajsad* (Book of the Middle Course concerning the Reformation of Souls and Bodies) gives a summary of diseases, therapeutics, and hygiene written specially for the benefit of the layman. Its initial part is a valuable discourse on psychology. *Kitab al-Aghthiya* (Book on Foodstuffs) describes different types of food and drugs and their effects on health.[83]

Ibn Zuhr, in his works, lays stress on observation and experiment, and his contributions greatly influenced medical science for several centuries, both in the East and the West. His books were translated into

Latin and Hebrew and remained popular in Europe as late as the advent of the eighteenth century.

Abu Abdallah Muhammad Ibn Muhammad Ibn Abdallah Ibn Idris al-Qurtubi al-Hasani (1100-1165)

Al-Idrisi was born in Ceuta, Spain and educated in Cordova. He is more famous for his studies and research on medicinal plants and geography. His major contribution lies in medicinal plants as presented in his several books, especially *Kitab al-Jami-li-Sifat Ashtat al-Nabatat*. He studied and reviewed all the literature on the subject of medicinal plants and formed the opinion that very little original material had been added to this branch of knowledge since the early Greek works. He, therefore, collected plants and data not reported earlier and added this to the subject of botany, with special reference to medicinal plants. Thus, a large number of new drug plants together with their evaluation became available to the medical practitioners. He has given the names of the drugs in six languages: Syriac, Greek, Persian, Hindi, Latin and Berber.[84]

In addition to the above, he made original contributions to geography, especially as related to economics, physical factors, and cultural aspects. At the request of the Norman King of Sicily, Roger II, he wrote a famous book on geography entitled *Nuzhat al-Mushtaq fi'I khtiraq al-Afaq* (The delight of him who desires to journey through the climates). This is practically a geographical encyclopaedia of the time, containing information not only on Asia and Africa but also Western countries. He produced an atlas of 70 maps of various regions called *Al-Kitab al-Rujari* (Roger's Book), also showing that the Earth was round, which was a common notion held by Muslim scholars. Al-Idrisi drew the global map centuries before Marco Polo or Columbus.[85]

Al-Idrisi, later on, also compiled another geographical encyclopaedia, larger than the former, entitled *Rawd-Unnas wa Nuzhat al-Nafs* (Pleasure of men and delight of souls), also known as *Kitab al-Mamalik wa al-Masalik*.[86] Apart from botany and geography, Idrisi also wrote on fauna, zoology, and therapeutical aspects. His work was soon translated into Latin and, especially, his books on geography, remained popular both in the East and the West for several centuries.

Abul Waleed Muhammad Ibn Ahmad Ibn Muhammad Ibn Rushd (1126-1198)

Ibn Rushd, known as Averroes in the West, was born in Cordova, where his father and grandfather had both been judges. He studied philosophy, law, logic, music, jurisprudence, mathematics, and medicine. He was a rationalist philosopher and was known as the best Islamic philosopher in the West. He wrote an extensive corpus of philosophy in which he stated significant theories in epistemology, natural philosophy, and metaphysics.

He was an accomplished physician, and on medicine alone he wrote 20 books, and the most famous treatise was *Kitab al-Kulyat fi al-Tibb*, known in Latin as the *Colliget*. In it, Ibn Rushd has thrown light on various aspects of medicine, including the diagnoses, cure and prevention of diseases. The book concentrates on specific areas in comparison to Ibn Sina's wider scope of Al-Qanun, but contains several original observations of Ibn Rushd. In the field of music, Ibn Rushd wrote a commentary on Aristotle's book *De Anima*. This book was translated into Latin by Mitchell Scott. In astronomy, he wrote a treatise on the motion of the sphere, *Kitab fi-Harakat al-Falak*. He also summarised Almagest and divided it into two parts: a description of the spheres, and the movement of the spheres. This summary of the Almagest was translated from Arabic into Hebrew by Jacob Anatoli in 1231. Regarding jurisprudence, his book *Bidayat al-Mujtahid wa Nihayat al-Muqtasid* has been held by Ibn Jafar Thahabi as possibly the best book on the Maliki School of Fiqh. Ibn Rushd's writings were translated into various languages, including Latin, English, German, and Hebrew. Ibn Rushd also wrote commentaries on Plato's Republic, Galen's treatise on fevers, Al-Farabi's logic etc. Eighty seven of his books have still survived.[87]

In philosophy, his most important work, *Tuhafut al-Tuhafut,* was written in response to Al-Ghazali's work. Ibn Rushd was criticised by many Muslim scholars for this book, which nevertheless had a profound influence on European thought, at least until the beginning of modern philosophy and experimental science. Like other rationalists before him, Ibn Rushd drew the ire of his opponents for suggesting that revelation must be guided by reason. In his opinion, the noblest form of worship was to study Allah through the Holy Quran, using the faculty of the mind. Ibn Rushd refuted the arguments of orthodox people or philosophers like Al-Ghazali that anything happens in Allah's will. Al-Ghazali holds that every time a fire burns cotton, the fire itself does not produce the burning effects; they are caused directly by Allah. It is in Allah's power to stop fire from producing these habitual effects, if and when He so wishes. Ibn Rushd says that the fire burns a piece of cotton not because of Allah's will but because of its roperties.[88]

Abu Yaqub, the Caliph of Morocco, called him to his capital and appointed him as his physician in place of Ibn Tufail. His son Yaqub al-Mansur retained him for some time, but soon Ibn Rushd's views on theology and philosophy drew the Caliph's wrath. All his books, barring strictly scientific ones, were burnt, and he was banished to Lucena. However, as a result of intervention of several leading scholars, he was forgiven after about four years and recalled to Morocco in 1198, but he died towards the end of the same year.[89]

Ibn Rushd has been held as one of the greatest thinkers and scientists and has influenced Western thought from the twelfth to the sixteenth centuries. His books were included in the syllabi of Paris and other universities until the advent of modern experimental sciences. Ibn Rushd was without question one of the greatest scholars and philosophers the Arab world has produced, and one of the profoundest commentators of Aristotle's works. In spite of his high position, he met a heavy setback in Caliph Abu Yusuf's reign. He was banned and his books were burnt.

Abu Muhammad Abdallah Ibn al-Baitar Dhiya al-Din al-Malaqi (1197-1248)

Ibn al-Baitar, born in the Spanish city of Malaqa (Malaga), was one of the greatest scientists of Muslim Spain and was the greatest botanist and pharmacist of the Middle Ages. He was a learned botanist, with whom he started collecting plants in and around Spain. He organised an expedition and travelled along the northern coast of Africa as far as Asia Minor. His visit included Bugia, Qastantunia (Constantinople), Tunis, Tripoli, Barqa, and Adalia. After 1224, he entered the service of al-Kamil, the Egyptian Governor, and was appointed chief herbalist. In 1227, al-Kamil extended his domination to Damascus, and Ibn al-Baitar accompanied him there, which provided him an opportunity to collect plants in Syria. His research on plants extended over a vast area, including Arabia and Palestine, which he either visited or managed to collect plants from stations located there.[90]

Ibn Baitar's major contribution, *Kitab al-Jami fi al-Adwiya al-Mufrada*, is one of the greatest botanical compilations dealing with medicinal plants in Arabic. The encyclopaedia comprises some 1400 different items, largely medicinal plants and vegetables, of which about 200 plants were not known earlier. The book refers to the work of approximately 150 authors, mostly Arabic, and quotes approximately 20 early Greek scientists. It was translated into Latin and published in 1758. His second monumental treatise, *Kitab al-Mlughni fi al-Adwiya al-Mufrada,* is an encyclopaedia of medicine. The drugs are listed in accordance with their therapeutical value. Thus, its 20 different chapters deal with the plants bearing significance to diseases of the head, ear, eye etc. On surgical issues, he has frequently quoted the famous Muslim surgeon, Abul Qasim Zahrawi. Besides Arabic, Baitar has given Greek and Latin names of the plants, thus facilitating the transfer of knowledge.[91]

Ibn Baitar's contributions are characterised by observation, analysis, and classification and have exerted a profound influence on Eastern as well as Western botany and medicine. Though the Jami was translated and published late in the Western languages, as mentioned above, many scientists had earlier studied various parts of the book and made several references to it.

Yaqut Al-Hamawi (1179-1229)

Al-Hamawi, born in Constantinople, was a great geographer, biographer, and Islamic historian of the twelfth and thirteenth centuries. Of Greek parentage, he was taken prisoner as a young boy and brought to Bagdad, where he accepted Islam, learned Arabic, and spent the rest of his life travelling throughout the Muslim world. A man of outstanding learning, Yaqut left us numerous encyclopaedic works, four of which have been discovered. Among them is his great geography, *Muajam al-Buldan*, an encyclopaedia of Islam, which has achieved the status of a classic and is still used as a reference work by scholars in the Muslim world as well as in the West.[92]

Arranged in alphabetical order, the *Muajam* has preserved a wealth of information not only on geographical positions, boundaries, and coordinates but also on scholars, artists, and scientists. Passionately given to detail, Yaqut's geographical work is intimately linked with history; he was equally concerned with correct orthography, because he was aware that slight sloppiness could lead to big blunders. His inspiration to compile such a geographical dictionary came from the Quran, as he himself writes in his introduction.[93]

Abu Jafar Muhammad Ibn Muhammad Ibn al-Hasan Nasir al-Din al-Tusi (1201-1274)

Al-Tusi, born in Tus (Khurasan), was known by a number of different names during his lifetime, such as Muhaqqiq-i Tusi, Khwaja-yi Tusi, and Khwaja Nasir. In Tus, al-Tusi's father was a jurist in the Twelfth Imam School. The Twelfth Imam was the main sect of Shi'ite Muslims, and the school, where Al-Tusi was educated, was mainly a religious establishment. From Tus, al-Tusi went to Nishapur, which was a good choice for Al-Tusi to complete his education since it was an important centre of learning. Al-Tusi studied logic, physics, metaphysics, philosophy, medicine, and mathematics.

Al-Tusi was invited by the Isma'ili ruler Nasir ad-Din Abd ar-Rahim to join the service of the Assassins, and was appointed a highly regarded member of the Isma'ili Court. During this period, he wrote important works on logic and philosophy, as well on mathematics and astronomy. The first of these works, *Akhlaq-i Nasiri*, was written in 1232. It was a work on ethics which Al-Tusi dedicated to the Isma'ili ruler Nasir ad-Din Abd ar-Rahim.[94]

In 1256, Al-Tusi joined Hulegu's service, which was deeply impressed by his knowledge, including his astrological competency. Hulegu appointed him as one of his ministers, scientific adviser, in charge of religious affairs, and administrator of Auqaf. When al-Tusi presented Hulegu with plans for the construction of a fine observatory at his capital, Maragheh, Hulegu was happy to agree. Maragheh was in the Azerbaijan region of northwestern Iran, and it was at Maragheh that the observatory was to be built. Construction of the observatory began in 1259 and became operational in 1262. Interestingly, the Persians were assisted by Chinese astronomers in the construction and operation of the observatory. It had various instruments, such as a 4 metre wall quadrant made from copper and an azimuth quadrant, which was the invention of Al-Tusi himself. Al-Tusi also designed other instruments for the observatory which was far more than a centre for astronomy. It possessed a fine library with books on a wide range of scientific topics, while work in science, mathematics, and philosophy were vigorously pursued there.[95]

Al-Tusi put his observatory to good use, making very accurate tables of planetary movements. After 12 years of observation, he published Zij-i-Ilkhani (the Ilkhanic Tables). This work contains tables for computing the positions of the planets, and it also contains a star catalogue. This was not the only important work that Al-Tusi produced in astronomy. It is fair to say that Al-Tusi made the most significant development of Ptolemy's model of the planetary system up to the development of the heliocentric model in the time of Copernicus.[96]

Al-Tusi put forward his concept known as the "Tusi Couple", a hypothetical model of "epicyclic motion that involves a combination of motions, each of which is uniform with respect to its own centre." Tusi Couple resolves linear motion into the sum of two circular motions. The aim of Al-Tusi, with this result, was to remove all parts of Ptolemy's system that were not based on the principle of uniform circular motion. He also wrote several works on astronomical instruments, for example, on constructing and using an astrolabe.[97]

Ptolemy's *Almagest* was one of the works which Arabic scientists studied intently. In 1247, Al-Tusi wrote *Tahrir al-Majisti* (Commentary on the *Almagest*), in which he introduced various trigonometrical techniques to calculate tables of sines. As in the *Zij-i Ilkhahi,* Al-Tusi gave tables of sines with entries calculated to three sexagesimal places for each half degree of the argument.[98] .

One of Al-Tusi's most important mathematical contributions was the creation of trigonometry as a mathematical discipline in its own right rather than as just a tool for astronomical applications. In his Treatise on the quadrilateral, Al-Tusi gave the first extant exposition of the whole system of plane and spherical trigonometry. This work also contains the famous sine formula for plane triangles: a/sin A = b/sin B = c/sin C.[99] This work is really the first in history on trigonometry as an independent branch of pure mathematics and the first in which all six cases for a right-angled spherical triangle are set forth.

He also wrote a famous work on minerals, which contained an interesting theory of colour based on mixtures of black and white, and included chapters on jewels and perfumes. He also wrote on medicine, but his medical writings are among his least important. Much more important were Al-Tusi's contributions to philosophy and ethics. In particular, in philosophy, he asked important questions about the nature of space.

Ala-al-Din Abu al-Hasan Ali Ibn Abi al-Hazm al-Qarshi al-Damashqi al-Misri (1213-1288)

Ibn al-Nafis, as he is popularly known, was born in Damascus. After acquiring his expertise in medicine and jurisprudence, he moved to Cairo, where he was appointed as the Principal at the famous Nasri Hospital. Here he imparted training to a large number of medical specialists, including Ibn al-Quff al-Masihi, the famous surgeon. He also served at the Mansuriya School in Cairo. When he died, he donated his house, library, and clinic to the Mansuriya Hospital. Apart from medicine, Ibn al-Nafis learnt jurisprudence, literature, and theology. He thus became a renowned expert on the Shafi'i School of Jurisprudence as well as a reputed physician.

His major contribution lies in medicine. His method included writing detailed commentaries on early works, critically evaluating them, and adding his own unique contribution. His major original contribution of great significance was his discovery of the circulatory system of the blood, which was re-discovered by modern science after a lapse of three centuries. He advanced a theory of blood circulation between the compartments of the heart and the lungs, and of pulmonary circulation, or less circulation. He made several important contributions to the early knowledge of the pulmonary circulation of venous blood passing into the hearts and lungs *via* ventricles, thus becoming oxygenated and arterial blood. He elaborated on the function of the coronary arteries as feeding the cardiac muscle. He was the first to correctly describe the constitution of the lungs and gave a description of the bronchi and the interaction between the human body's vessels for air and blood.[100]

He was the first person to believe that all the blood that reached the left ventricle passed through the lung. He stated that there is no connection between the two cavities of the heart (right and left ventricles) and that blood cannot pass through the (interventricular) septum. He also stated that there must be small communications or pores between the pulmonary artery and vein, a prediction that predated by 400 years the discovery of the pulmonary capillaries by Marcello Malpighi.[101]

The most voluminous of his books is *Al-Shamil fi al-Tibb*, which was designed to be an encyclopaedia comprising 300 volumes, but it could not be completed due to his death. The manuscript is available in Damascus. His book on ophthalmology is largely an original contribution and is also extant. However, his book that became most famous was *Mujaz al-Qanun,* and a number of commentaries were written on it. His own commentaries include one on Hippocrates' book. He wrote several volumes on Ibn Sina's *Qanun* that are still extant. Likewise, he wrote a commentary on Hunayn Ibn Ishaq's book. Another famous book embodying his original contribution was on the effects of diet on health, entitled *Kitab al-Mukhtar fi al-Aghdhiya.*[102]

Ibn Al-Nafis' works integrated the then existing medical knowledge and enriched it, thus exerting a great influence on the development of medical science, both in the East and the West. However, only one of his books was translated into Latin at early stages and, therefore, a part of his work remained unknown to Europe for a long time.

Ibn abi al-Mahasin al-Halabi (1210-1267)

Al-Halabi, born in Aleppo, was an eminent writer in the field of eye surgery. He was named after Aleppo (Arabic: *Halab*), a city in the northern part of Syria, which was known for its thriving trade and wealth in the middle ages. He was considered the first ophthalmologist to use a magnet to remove metallic foreign bodies from the eye. He was aware of the concept that an instrument may be substituted by another instrument due to lack of availability. His surgical instruments represent the biomechanics field. Biomedical engineering students should know Al-Halabi for his numerous contributions in the surgical instruments field.

In his work "*Kitab al-Kafifi al-Kuhl*" (The Book of Sufficient Knowledge in Ophthalmology), he mentions eighteen major ophthalmologic texts. Al-Halabi's book was the first book to give a remarkable illustration of the anatomy of the brain, the eyes, and the visual pathway among them. In his work, he also describes and gives drawings of various surgical instruments, including 36 instruments for eye surgery. He was an intriguing ophthalmologist who invented many surgical instruments for the treatment of various eye diseases.[103] He included in his book a magnificent illustration of the anatomical structure of the eye. The book reflects Al-Halabi's medical practice and teaching, and shows a number of advanced medical techniques and tools. His invaluable comments reflect his deep experimental observations in the field of ophthalmology. His work was very practical, too, with very good descriptions of cataract operations and the instruments used, and also the steps to be taken after the operation.

Chapters in the treatise deal with eye diseases, their definitions, descriptions, varieties, causes, symptoms, treatments, medicines, and cures for them. Surgery for the removal of the cataract is described in good detail, including the author's own experience. Al-Halabi has given in his book the details of cataract operations representing twelve sorts of cataract. He dedicated a chapter of his book to the procedures for removing the cataract. He described cataract operations, the required instruments, and the steps to be taken after the operation. Al-Halabi's anatomy of the eye mentions: definition; colour of the eye; tunics of the eye; the fluids of the eye; visual power and its nerves; motive nerves; muscles of the eye, eyelids and eyelashes; hygiene of the eye; things which are useful and things which are harmful; how to open the eye and introduce drugs into it; the best kind of probe and its use; tools for the handling of ophthalmology; best garments for the eye doctor.[104]

In his work there are synoptic tables relative to the diseases of the eyes and eyelids, giving for each disease the definition, description, varieties, causes, symptoms, treatments, drugs, including narcotics; and other tables relative to surgical cases. Finally, there is a list of drugs. The most remarkable synoptic table is the one related to instruments; each table contains 18 such instruments, with their names and their usage. The instruments are in colour and perfectly drawn; some instruments are for the operations on cataracts, others for eye infections which do not affect the senses. The author himself is so confident in his own talents as an eye surgeon that he did not fear to operate the cataract of a one eyed man.

Ibn al-Shatir (1304-1375)

Ibn al-Shatir was an Arab astronomer, mathematician, and engineer. Ibn al-Shatir applied the epicyclical motion model of Al-Tusi to the motions of the heavenly bodies in the fourteenth century. Ibn al-Shatir's formulations were the beginnings of verifying theoretical astronomy through systematic observations. Ibn al-Shatir's theory of lunar motion was very similar to that attributed to Copernicus some 150 years later.

Currently, researchers are investigating whether it was possible that Copernicus, while visiting the Vatican library in Rome, had seen Ibn al-Shatir's fourteenth century manuscript illustrating his concept of planetary motion. The reason for this supposition was a diagram in Copernicus' Commentaries, which was remarkable and similar to Ibn al-Shatir's schematic diagrams. Whereas Ibn al-Shatir's concept of planetary motion was conceived in order to play an important role in an earth-centred planetary model, Copernicus used the same concept of motion to present his sun-centred planetary model. Thus, the development of alternative models took place that permitted empirical testing of the models.[105]

Abd al-Rahman Ibn Mohammad Ibn Khaldun (1332-1406)

Abd al-Rahman Ibn Mohammad, generally known as Ibn Khaldun, was born in Tunisia, where he received his early education and where, still in his teens, he entered the service of the Egyptian ruler Sultan Barquq. His thirst for advanced knowledge and a better academic setting soon made him leave this service and migrate to Fez.

In this turbulent period, he took refuge in a small village, Qalat Ibn Salama, in Algeria for three years, which provided him with the opportunity to write *Muqaddimah*, the first volume of his world history that won him an immortal place among historians, sociologists, and philosophers. Ibn Khaldun then went to Egypt and here he lived a life of fame and respect, marked by his appointment as the Chief Malakite Judge and lecturing at Al-Azhar University, but envy caused his removal from his high judicial office as many as five times.[106]

Ibn Khaldun's chief contributions lie in the philosophy of history and sociology. He sought to write a world history prefaced by a first volume aimed at an analysis of historical events. This volume, commonly known as *Muqaddimah* or 'Prolegomena', was based on Ibn Khaldun's unique approach and original contribution and became a masterpiece in literature on the philosophy of history and sociology. The chief concern of this monumental work was to identify psychological, economic, environmental, and social facts that contribute to the advancement of human civilization and the currents of history. In this context, he analysed the dynamics of group relationships and showed how group-feelings, *Al-Asabiyya*, give rise to the ascent of a new civilisation and political power and how, later on, its diffusion into a more general civilization invites the advent of a new *Asabiyya* in its pristine form. He identified an almost rhythmic repetition of rise and fall in human civilization, and analysed the factors contributing to it.[107]

Apart from the *Muqaddimah* that became an important independent book even during the lifetime of the author, the other volumes of his world history *Kitab al-I'bar* deal with the history of Arabs, contemporary Muslim rulers, contemporary European rulers, ancient history of Arabs, Jews, Greeks, Romans, Persians etc., Islamic history, Egyptian history, and North-African history, especially that of Berbers and tribes living in the adjoining areas. The last volume deals largely with the events of his own life and is known as *Al-Tasrif*. This was also written in a scientific manner and initiated a new analytical tradition in the art of writing autobiography. A book on mathematics written by him is not extant.[108]

Ibn Khaldun's contribution to history is marked by the fact that, unlike earlier writers interpreting history largely in a political context, he emphasised environmental, sociological, psychological, and economic factors governing the apparent events. This revolutionised the science of history and also laid the foundations of Umraniyat (Sociology). His influence on the subjects of history, philosophy of history, sociology, political science, and education has remained paramount ever since his life. His books have been translated into many languages, both in the East and the West, and have inspired subsequent development of these sciences.

Serafeddin Sabuncuoglu (1385–1468)

Serafeddin Sabuncuoglu was a mediaeval Ottoman surgeon and physician. He started and continued his medical interests and practice in Amasya Dar-es Sifa Hospital until he died. Serafeddin Sabuncuoglu was also the director of the Amasya Dar-es Sifa Hospital for almost 14 years. Serafeddin Sabuncuoglu was the author of the *Cerrahiyyetu'l-Haniyye* (Imperial Surgery), the first illustrated Turkish-written medical textbook. Sabuncuoglu's *Imperial Surgery* was the first surgical atlas and the last major medical encyclopaedia from the Islamic world. Female surgeons were also illustrated for the first time in *Imperial Surgery*. One of the surgical techniques, described by Sabuncuoglu, was the ligating of the temporal artery for migraine.[109]

These include three chapters that incorporate 134 surgical interventions and 156 surgical instruments. These three chapters discussed cauterization treatments, surgical procedures, fractures and dislocations, and cancer-related issues. Serafeddin Sabuncuoglu was one of the first to describe hydrocephalic drainage techniques in children. Hydrocephalus is a medical condition in which fluid builds up in the brain. Sabuncuoglu was also the first to use an axial traction technique for spinal surgery. An important observation that Serafeddin Sabuncuoglu made in *Imperial Surgery* was that a fast delivery of a baby can occur if the patient holds their breath and exerts a strong force.[110]

Another book of Serafeddin Sabuncuoglu was the *Mucerrebname* (The book of tried medications). It was the manual on drugs and their preparation. This was the last of three medical treatises he composed. He explains, following common phrasing we see in many medical treatises, that he wrote the work as an answer to a plea from colleagues that he put on paper his fourteen-year medical experiences in a hospital. The contents and structure of this short manual—with regard to types of drugs, their ingredients, and methods of preparation—repeat earlier Arab-Islamic examples of learned written pharmacopeia. At the same time he refers frequently to his techniques in preparing drugs and discusses efficacy based on experiments upon himself and animals.[111]

NOTES

1. Hoodbhoy, Pervez (1991). *Islam and Science: Religious Orthodoxy and the Battle for Rationality*, London, p. 96.
2. Two major internal fissures in the Muslim polity were also great obstacles to scientific and rational development. The first (656-661) was sparked by the assassination of the third Caliph, Othman, and led to a civil war in which close Companions of the Prophet found themselves pitched against each other under circumstances which threatened the very existence of the Muslim community. During the second rift (680-692), which sprang from two rival claims to the Caliphate, Husain Ibn Ali, the grandson of the Prophet, and all but one of his companions, were killed at Karbala in October 680; Makkah was besieged by armed men; a radical splinter group, the Khawarij, took control of much of Arabia; and Ibn al-Zubair, one of the close companions of the Prophet, was killed in the sanctified city of Makkah, where fighting had been declared unlawful by the Quran. [Muzaffar Iqbal (2009). *The Making of Islamic Science*, Kuala Lumpur, p. 14.]
3. Nasr, *op. cit.*, p. 11.
4. Walbridge, *op. cit.*, p. 56.
5. Iqbal, Muzaffar, *op. cit.*, p. 28.

6. *Ibid.*
7. *Ibid.*, p. 17.
8. *Ibid.*, pp. 28-29.
9. Hoodbhoy, *op. cit.*, pp. 90-91.
10. Iqbal, Muzaffar, *op. cit.*, pp. 17-18.
11. Hoodbhoy, *op. cit.*, p. 91.
12. *Ibid.*, p. 97.
13. *Ibid.*, pp. 98-99; see also Salim T.S. Al-Hassani (3rd Edition, 2012). *1001 Inventions: The Enduring Legacy of Muslim Civilization*, Washington DC, p. 72.
14. Wan Hazmy C. H., Zainurashid Z., Hussaini R. (2009). *Biography: Muslim Scholars and Scientists*, N. Sembilan. pp. 5-6.
15. Iqbal, Muzaffar, *op. cit.*, pp. 15-16; see also Seyyed Hossein Nasr (2001). *Science and Civilization in Islam*, Chicago, p. 43.
16. Wan Hazmy, *op. cit.*, p. 6.
17. Al-Hassani, *op. cit.*, p. 92.
18. Wan Hazmy, *op. cit.*, p. 7.
19. *Ibid.*, p. 10.
20. Around the year 825, Al-Khwarizmi with the encouragement of Caliph Al-Mamun wrote *Al-jabr wa'l-Muqabalah*, which Khwarizmi himself writes that it would be useful in the calculation of what men constantly need to calculate for their inheritance and legacies, their portions and judgments, in their trade and in all their dealings with one another in matters involving measurement of land, the digging of canals, and geometrical calculations, and other matters involving their crafts. Three centuries later, this book was partially translated into Latin by Robert of Chester as *Liber Algebras et Almucabola*; shortly afterwards, Gerard of Cremona retranslated it as *De Jebra et Almucabola*, and introduced into Europe a science completely unknown till then, and with it, a terminology which was still capable of growth but already completely developed. In this manner, the title of the young man's book inaugurated a new branch of science, algebra. Iqbal, Muzaffar, *op. cit.*, pp. 48-49.
21. Kettani, M.A. (1976). *Moslem contributions to the natural sciences*. Impact of Science on Society, 26(3), 135-147. Al-Khwarizmi in the introduction to his Algebra, he wrote, "I confined this book to what men constantly need to calculate their inheritance and legacies, their portions and judgments, in their trade and in all their dealings with one another in matters involving measurement of land, the digging of canals, and geometrical calculations, and other matters involving their crafts." (Iqbal, Muzaffar, *op. cit.*, p. 74.)
22. Wan Hazmy, *op. cit.*, pp. 11-12; see also Beshore, G. (1998). *Science in Early Islamic Culture*. New York, p. 24.
23. *Ibid.*, p. 11.
24. *Ibid.*, pp. 13-14; see also *Science and Civilization in Islam*, *op. cit.*, p. 44.
25. *Ibid.*, p. 14.
26. *Ibid.*, p. 18.

27. Hoodbhoy, *op. cit.*, p. 110. Al-Kindi rationalised his efforts at allegorical interpretation and gives the example of Verse 22 of Surah Al-Hajj of the Quran. In this passage, the believer is told that the sun, moon, stars, mountains, trees, and beasts 'bow themselves' before God (*Al-Quran*, 22:18). For the unsophisticated, this invokes an image wherein all creation literally bends in prayer. But Al-Kindi gave an elaborate linguistic argument that the Arabic word for 'bow' should be understood as meaning 'obey'. Thus, the naive picture of universal worship should instead be understood as the universal obedience to God's will.
28. *Ibid.*
29. *Ibid.*, p. 111. For his rational approach to the Islamic philosophy, the sixty-year-old Al-Kindi received fifty lashes before a large crowd which had assembled. Long before his death in 873 at the age of seventy-two, Al-Kindi had succumbed to prolonged depression and silence. Al-Kindi was the first major figure of Islamic scholarship to fall victim to the orthodox reaction against rationalism.
30. Wan Hazmy, *op. cit.*, pp. 21-22.
31. *Ibid.*, p. 23.
32. *Ibid.*, pp. 25-26.
33. *Ibid.*, p. 26.
34. *Ibid.*, pp. 29-30.
35. *Ibid.*, p. 30.
36. *Ibid.*, pp. 30-32.
37. *Ibid.*, p. 39.
38. *Ibid.*, p. 40.
39. Faruqi, Yasmeen Mahnaz, "Contributions of Islamic scholars to the scientific enterprise", *International Education Journal*, 2006, 7(4), pp. 391-399.
40. Wan Hazmy, *op. cit.*, p. 43.
41. *Ibid.*, pp. 43-44.
42. *Ibid.*, p. 44.
43. Hoodbhoy, *op. cit.*, pp. 111-12.
44. Wan Hazmy, *op. cit.*, pp. 34-35.
45. *Ibid.*, pp. 35-36; see also Kettani, *op. cit.*, 26(3), 135-147.
46. *Ibid.*, pp. 36-37.
47. *Ibid.*, p. 46.
48. *Ibid.*, pp. 46-47; see also Ameer Ali, *op. cit.*, p. 426.
49. *Ibid.*, p. 48.
50. *Ibid.*, p. 49; see also Bhargava, *op. cit.*, pp. 233-34.
51. Chaney, Eric (2008). *Tolerance, Religious Competition and the Rise and Fall of Muslim Science*. Harvard University, pp. 15-37.
52. *Ibid.* see also Wan Hazmy, *op. cit.*, p. 57.
53. Wan Hazmy, *op. cit.*, p. 52.
54. *Ibid.*, p. 53.
55. Al-Hassani, *op. cit.*, p. 54.

56. *Ibid.*
57. Wan Hazmy, *op. cit.*, p. 59.
58. Ameer Ali, *op. cit.*, pp. 377-78.
59. Dorsey Armstrong (2014). *Great Minds of the Mediaeval World*, Virginia, p. 56; see also Al-Hassani, *op. cit.*, pp. 55-56. (One day, he saw light shining through a pinhole into his darkened room, projecting an image of the world outside onto the opposite wall. At the time, people believed the ancient view that our eyes see by sending out invisible rays. But Al-Haitham realised that we see because light rays emitted from visible objects actually enter our eyes. By experimenting with a dark room "camera obscura" he showed that the light enters our eyes in the shape of a geometric cone of vision.)
60. Hoodbhoy, *op. cit.*, p. 112.
61. Wan Hazmy, *op. cit.*, p. 69; see also Faruqi, Yasmeen Mahnaz, "Contributions of Islamic scholars to the scientific enterprise", *International Education Journal*, 2006, 7(4), 391-399.
62. Faruqi, *op. cit.*, 391-399. (The first true medical schools of the Middle Ages were founded in Salerno, Italy; there, starting in the 9th century, ancient traditions of Greek and Latin healing were taught according to a well-organised curriculum, and medical treatises from the North African and Muslim worlds were incorporated from an early date. But it was when Avicenna's Canon of Medicine arrived in the 12th century that medicine as a subject began to be codified and standardised throughout the mediaeval world. Eventually, medical schools opened in Bologna, Padua, Paris, Oxford, and other European cities. And in all of them, Avicenna's Canon of Medicine was the basis of education. Dorsey Armstrong, *op. cit.*, p. 45.)
63. Wan Hazmy, *op. cit.*, p. 70; see also Dorsey Armstrong, *op. cit.*, p. 46.
64. *Ibid.* see also Hunt Janin (2005). *The Pursuit of learning in the Islamic World 610-2003*, McFarland & Co Inc. USA, p. 80.
65. *Ibid.,* p. 71.
66. Hoodbhoy, *op. cit.*, p. 112.
67. *Ibid.* (For Avieenna, God knows nothing about human history. God understands the world of which He is the first cause, but He does not (nor should He) know about individuals or particulars. Avicenna argues that even evil things are "accidental" consequences of what are ultimately good things. This position caused consternation for later Muslim scholars, who saw it as a heretical denial of God's intervention in and knowledge of the world he had created. Dorsey Armstrong *op. cit.*, p. 50.)
68. Wan Hazmy, *op. cit.*, p. 64.
69. *Ibid.*
70. Iqbal, Muzaffar, *op. cit.*, p. 66.
71. Wan Hazmy, *op. cit.*, p. 64.
72. Afridi, Mohammad Adil (2013). *Contribution of Muslim Scientists to the World: An Overview of Some Selected Fields*, Revelation and Science, (*Inter-Disciplinary Journal of Intellectual Revival*, International Islamic University, Pahang, Malaysia) Vol. 3, No. 1, 2013. pp. 50-51.

73. Ahmad, Imam Ibrahim (1967 Eng. Trans.). *Kitab Tahdid al-Amakin* of Al-Biruni, Beirut, p. 16; see also Wan Hazmy, *op. cit.*, p. 65.
74. Iqbal, Muzaffar, *op. cit.*, p. 107. (There is a story that when Al-Biruni was dying, his friend came to see him. Al-Biruni enquired about the solution of some mathematical problem regarding which that friend had spoken to him earlier. The friend was shocked that he was worrying about such things in that condition. Al-Biruni replied with a great effort and wanted to know if it was not desirable that he should die with the knowledge of the solution to that problem rather than without it. The friend gave him the desired information, and as he walked out of the room, he heard people crying at the death of Al-Biruni. Ahmad, Qeyamuddin (1988). *India by Al-Biruni*, New Delhi, p. XVII.)
75. Umar Khayyam himself said the difficulties, "I was unable to devote myself to the learning of this algebra and the continued concentration upon it, because of obstacles in the vagaries of time which hindered me; for we have been deprived of all the people of knowledge save for a group, small in number, with many troubles, whose concern in life is to snatch the opportunity, when time is asleep, to devote themselves meanwhile to the investigation and perfection of a science; for the majority of people who imitate philosophers, they confuse the true with the false, and they do nothing but deceive and pretend knowledge, and they do not use what they know of the sciences except for base and material purposes; and if they see a certain person seeking for the right and preferring the truth, doing his best to refute the false and untrue and leaving aside hypocrisy and deceit, they make a fool of him and mock him." Wan Hazmy, *op. cit.*, pp. 77-78.
76. Iqbal, Muzaffar, *op. cit.*, p. 52.
77. Wan Hazmy, *op. cit.*, p. 78.
78. *Ibid.*, p. 79.
79. *Ibid.*, p. 83.
80. Ameer Ali, *op. cit.*, p. 428.
81. *Ibid.*, p. 429.
82. Wan Hazmy, *op. cit.*, p. 101.
83. *Ibid.*, p. 102.
84. *Ibid.*, pp. 103-104.
85. Iqbal, Muzaffar, *op. cit.*, p. 45.
86. Wan Hazmy, *op. cit.*, p. 104.
87. *Ibid.*, pp. 106-107.
88. Hoodbhoy, *op. cit.*, p. 114.
89. Wan Hazmy, *op. cit.*, p. 105.
90. *Ibid.*, p. 109.
91. *Ibid.*, p. 110.
92. Iqbal, Muzaffar, *op. cit.*, p. 45.
93. *Ibid.*
94. Wan Hazmy, *op. cit.*, p. 113.

95. *Ibid.*, p. 114.
96. Turner, R.H. (1995). *Science in Mediaeval Islam: An Illustrated Introduction*. Austin, p. 68.
97. *Ibid.*
98. Wan Hazmy, *op. cit.*, p. 116.
99. *Ibid.*
100. *Ibid.*, pp. 122-23.
101. John B. West, *Ibn al-Nafis, the pulmonary circulation, and the Islamic Golden Age*, J Appl Physiol (1985). 2008 Dec.; 105(6): 1877-1880.
102. Wan Hazmy, *op. cit.*, p. 123.
103. *https://muslimheritage.com/al-mahassin-al-urdi-al-lubudi-al-halabi/*and *https://www.researchgate.net/publication/331272985_Halabi.*
104. *Ibid.*
105. Sabra, A. I. (Jul./Aug. 2002). "Greek astronomy and the Mediaeval Arabic tradition". *American Scientist*, 90(4), 360-397; see also Saliba, G. (Jul./Aug. 2002). "Greek astronomy and the Mediaeval Arabic tradition". *American Scientist*, 90(4), 360-367.
106. Wan Hazmy, op. cit., p. 124.
107. *Ibid.*, p. 125.
108. *Ibid.*
109. Miri Shefer-Mossensohn (2015). *Science among the Ottomans: The Cultural Creation and Exchange of Knowledge*, Austin, pp. 31-32.
110. *Ibid.*
111. *Ibid.*

7
Trade and Industry in Mediaeval Islamic World

Before the Muslim civilization collapsed, it was able to register remarkable achievements. Its legacy is visible in great architectural works, modern sciences and learning, trigonometry, surgery, optics, the Arabic numerals, star names, trade and industry, the urban character of Islamic society, Islamic gardens and gardening, general aspects of Islamic scholarship, hospital care, university learning, early knowledge of India, China and Scandinavia etc. It was not just Arabs or Muslims who were involved in this great legacy, but all ethnic groups and faiths. Islamic civilisation was the one and only truly universal civilisation that has ever existed in the history of humanity. The Islamic policy of inclusion went along with other manifestations of the same nature.

Paper Industry

One thing that arrived in Bagdad just in time to really help the translation movement, and the whole of Arabic scholarship, was paper. There is an apocryphal story that the Muslims learned the art of papermaking after the conquest of Samarkand in 704 when the Muslims came into contact with the Chinese and learnt the art of paper-making. From another thesis, it is said that the Muslims learnt papermaking from Chinese prisoners they caught at the Battle of Tallas in 751. It is also said that paper arrived from China with the many traders who were at that time journeying far across Asia, and that they brought back Chinese calligraphy as well as paper. Either way, it arrived in Islam just about the same time as the founding of Bagdad by the Abbasids. If in China, papermaking might have been an art from, in Bagdad it became an industry.[1]

Early Islamic paper-making involved many modern industrial processes, such as mass manufacturing, the use of machine power, the use of wind and water to activate such machinery, and industrial organization. In 706, Yusuf Amr of Makkah discovered the process of making paper from cotton. The first paper mill in Islam was set up in Bagdad in 793-95. After Bagdad, paper manufacturing rapidly spread West to Damascus, Tiberias, and Syrian Tripoli. The traveler, Nasir Ibn-Khusraw, said that Syrian paper was similar to that made in Samarkand but of better quality. High-quality Hamawi paper was used for important diplomatic and religious documents. In general, Syrian paper was regularly shipped to Egypt as it was to the Red Sea ports and India.[2]

Paper production reached Egypt in the late eighth century, and the first extant Muslim document made of paper dates from the year 796 and comes from that country. The earliest copy of the Quran on paper was recorded there in the tenth century. There, in Egypt, paper was largely produced at al-Fustat, Fuwwa, and Fayyum. From Egypt, paper making travelled further west, to North Africa, and reached Morocco, some time in the ninth or tenth century. From Morocco, paper production crossed the straight into Muslim Spain; the Andalusians, according to the geographer Al-Muqaddasi, soon excelling at the craft.[3]

The town of Jativa, near Valencia, became famed for its manufacture of thick, glossy paper, the Shatihi, following a process borrowed from the East. According to the Sicilian-based Muslim geographer Al-Idrisi, the paper of Jativa was of such good quality that the like of it was not to be found anywhere else in the world, and that it was exported to the East and the West. There were also other places of paper production of the heavy, smooth, glazed sort. By the year 1000, it could be said that paper was in general use throughout the Islamic world, and this had a considerable impact on a wide variety of activities and industries. The advent of cheap and abundant paper fostered cultural activities and the industry of book production in the Islamic world. Markets for book makers and bookshops were spread in the vicinity of the mosques. Bagdad, most probably, became the first major city where the Warraqeen (paper maker) bookshops appeared. From there, these shops spread to other parts of the Muslim world; Cairo, most particularly, became a major centre of the Warraqeen.[4]

As early as the late eighth century, paper was introduced to government offices by Harun al-Rashid's Vizier Jafar. Good-quality papers were also used as gifts for sultans or as a material for currency; and as early as the nineth century, coloured papers were in use for a diversity of purposes—yellow, for instance, was favoured in religious circles as it facilitated the memorization of the Quran. Paper was also used for wrapping goods; in 1035, Nasir Ibn-Khusraw, on a visit to Cairo, expressed his astonishment at seeing—'sellers of vegetables, spices, and hardware, provided with paper on which all they sold was immediately wrapped up.'[5]

The paper industry stimulated a variety of crafts and professions, which included dying, ink making, manuscript craftsmanship, calligraphy etc. A number of sciences also progressed in the wake of the paper industry. Chemistry, obviously, has the most intricate links with the paper industry. They deal with inks, glues, and erasure fluids. Ibn Badis (1007-1061), a Tunisian, writes in his Umdat al-Kuttab, in twelve chapters, about the excellence of the pen, the preparation of different types of inks, the preparation of coloured inks, the colouring of dyes and mixtures, secret writing, the making of paper, and so on.[6]

M. Lombard in *Les Textiles*; and D. Hunter in *Paper-making*; provides a very informative outline on the manner Muslims manufactured paper. Hunter explains that Muslims disintegrated linen rags, placing them in heaps, saturating them with water, and allowing fermentation to take place, then boiled the mass in wood ash. The boiled rags were then placed in cloth bags and suspended in a running stream, the action of which removed a great part of the alkaline residue and much of the dirt.... Later, to beat the cleansed rags to a pulp, a trip hammer was put to use.[7]

There is nothing pioneering on the part of the Muslims in the invention of paper itself; a Chinese invention, like much else; the pioneering Muslim role, however, was in other ways. The Muslims adopted paper making for their specific local conditions, using local materials. The Chinese relied on fibre from the bark of the paper mulberry tree to prepare their material; but being short of the substance, the Muslims innovated, using primarily linen and cotton as a substitute. In Muslim Spain, generally, paper was manufactured from flax, linen, and cotton rags.

Textile Industry

The textile industry was widespread throughout the Islamic world. Mediaeval Muslim geographers provide excellent information on the numbers of looms, the craftsmen employed in the trade, and the burgeoning suburbs gravitating around textile manufacturing centres. In Tinnis, in the Egyptian Delta, 5,000 looms were recorded; 135 in Fustat, 900 could be found; 4000 in Tripoli, Syria; in Spain, in Almeria, 800 silk weaving looms were recorded, and so on and so forth.[8]

Al-Idrisi counted as many as three thousand farms raising silk worms in the mountains around Jaen city in Spain, in the middle of the twelfth century, and Al-Shaqundi labelled this region 'Jaen of the silk' because so many people, both in the city and countryside, were involved in the silk. Like Spain, for silk, Morocco was famous for cotton production. In Morocco, the irrigated high plains of Dai and Tadla and the Um Rabi Valley produced considerable quantities of cotton. This made the country an extremely important centre for cotton manufacturing, and a large share of the manufactured products were exported to the Sudan.[9]

Syria was a particularly important industrial centre; cotton weaving workshops in particular, al-Muqaddasi referred to the cloth manufacturing workshops of Aleppo. From Manbig and Aleppo, the cotton of Wadi Butnan was conveyed to the manufacturing centres of eastern Mesopotamia and western Syria. It seems that the Syrian workshops specialised in the production of cloth made of a mixture of silk and cotton; Syria, after all, was the largest producer of both items in the whole Mediterranean Basin. Besides quantity, quality was dominant too. Al-Muqaddasi notes the excellent quality of the close-woven garments produced on the island of Sicily.[10]

There were dominant aspects of ownership and management of the textile industry. Every Muslim ruler had at his disposal a workshop for the manufacture of his preferred cloth. Besides this, many privately owned workshops worked towards meeting the needs of the population and for export abroad. In Egypt, throughout the early Middle Ages, textile manufacture was partly a state-owned industry, but there were also private capitalists employing wage workers for their own benefit, at very low wages, and under the fairly strict supervision of the state. The same was the case in the large urban working communities in Almeria and Meknes in Morocco. In the eastern half of Islam, large silk weaving establishments existed in Tyre, Antioch, Tripoli, and other cities, four thousand weavers being engaged in this craft in Tripoli alone.[11]

Islamic textile products have found a large market stretching over the continents. The coloured garments produced in Basra, near Fes, in Morocco, possibly derived from both linen and cotton, were very much appreciated further south on the African continent, and merchants from the oasis of Sijilmasa (a mediaeval Islamic Moroccan city) traded such items in exchange for gold. Andalusi fabrics made their way to the Christian world as early as the ninth century. From the East, Syrian silk cloth was a popular export product, and was regarded as equalling the Chinese in quality. Syria, in the Middle Ages, was, thus, the largest supplier of cotton in the whole Mediterranean world, and its products were also renowned for their quality.[12]

The Islamic textile industry had a significant impact on other industries and the rise of technological expertise, both in the Islamic world and later in Western Christianity. The cotton industry, for instance, acted on the farming sector and stimulated the rise of dyestuffs, such as indigo and other fibres, such as linen, which were often mixed with cotton. Cotton also acted on the manufacture of clothing and household articles made of cotton cloth or cotton wool. In respect to stimulating technological innovations, textile machinery was already in existence in Muslim Spain centuries before it appeared in the rest of Western Christendom. The early phase, lasting from 825 to 925, was marked by two interesting technological innovations, such as the horizontal loom. So the horizontal loom was already in use in Al-Andalus at least three centuries before the rest of Europe, giving rise to a weaving industry.

Glass Industry

An important thing must be known before throwing light on the glass industry: it is not Islamic civilisation that started glass making, or many of the techniques involved with the craft. However, just as with paper, Islamic civilisation innovated in many respects, on both the qualitative (aspects of which have been just seen) and quantitative level, i.e. the Muslims promoted the craft into an industry or pre-industry, involving large number of workers, often working in association. It could not be denied that there was a very strong link between the faith of Islam and the rapid expansion of the glass industry. The Muslims transformed their mosques, houses, and cities into centres of rich decoration, but in this atmosphere, a special significance was accorded to glass, just as ceramics.

Abundant accounts from the whole mediaeval period testify to a thriving glass industry, which produced glasses of all shapes and colours, and was established in the cities of Alexandria, Cairo, Tyre, Aleppo, Antioch, Damascus, Tripoli, and elsewhere. The lamp shades were made of glass adorned with medallions, inscriptions, or floral designs. Syria was most particularly famous for the art of enamelled glass, which would reach its peak of excellence in the thirteenth century. Both Aleppo and Damascus were equally important centres, and so was Hama, where excavations yielded a wide range of material, mostly from the period between 1100 and 1400. Egypt also continued to produce vessels of all qualities, particularly at Al-Fustat (old Cairo), where excavations yielded immense quantities of glass, ranging in date from the eighth century to the late Middle Ages.[13]

Advanced skills and techniques in glass production and decoration were common throughout the Islamic world. Besides wheel cutting, most of the techniques of surface manipulation were known in the Islamic world, including mould blowing and casting, stamped and pinched decoration, and applied ornament. Under the Tulunids, Abbasids, Fatimids, Ayyubids, and

Mamluks, Muslim glass-makers created a new art of decorated glass hardly surpassed at any earlier or later time for strength, design, and richness of colour. The glass workers of the early Islamic period experimented with polychrome decoration of the surface; opaque coloured vessels were made in Egypt and incrustation with glass on another colour, the so-called technique of 'combed' decoration, was practiced in Syria.[14]

Some of the most sophisticated Egyptian glass vessels were decorated with lustre, a shiny, sometimes metallic effect, achieved by painting copper or silver oxide on the surface of the object, then firing it at a temperature of about 600°C in reducing conditions. In Spain, Ibn Fimas introduced the art of glass manufacturing and taught artisans the technique of putting the clay in the oven. Ibn Fimas, according to the sixteenth century historian Al-Maqqari, established factories and was the first to make glass out of clay. He advised the glassmakers to keep the crystal in the oven for twenty-four hours to make it strong.[15]

The Seljuks acted as patrons of many aspects of Islamic learning and crafts, and they gave a good impetus to artistic life in the countries where they exerted their influence. Enamel painting on glass, just as the damascening of bronze, for instance, reached very high levels under the Seljuks before peaking under the Mamluks. The typical lamp form also originated in Syria at the time of the Seljuks. The export of Muslim glass, especially Syrian, was a dominant trait of the mediaeval economy. The gilded and enamelled glasses for which the Near East was renowned were, for the most part, the products of Syria, and these products were exported as far as Egypt, Mesopotamia, China, Russia, Sweden, and England.[16] Syrian glass had a widespread reputation, supplying the tastes and homes of wealthy Western households with a wide range of glass products in a variety of shapes and applications.

Sugar Industry

Sugar cane was initially cultivated in India, Indo-China, and also China proper, but neither the Chinese, nor the Indians, invented the manufacturing processes. For centuries, all they did was to crush the cane and extract the juices, which they then thickened on fire. However, these were the Muslims who invented sugar and established sugar mills in most parts of the Islamic Empire. In many Islamic parts, a strong sugar industry flourished, where modern processes of industrial production were in operation. In Islamic states, the sugar refineries were run not only by the state but also with the involvement and partnership of private manufacturers.

The method of refining sugar through chemical processes was used in Islamic lands; Muslims developed the art of refining sugar within the Caliphate and spread it beyond its borders. A thriving industry developed, and was located along the coastal area of the extreme east of the Mediterranean, in the centres of cultivation at Laodicea, Tripoli, Sidon, Tyre, Caesarea, Palestina, and Acre, whilst inland, the cane was established around Antioch, Aleppo, Damascus etc. In Damascus and Cairo, whole streets were set aside for sugar refineries. Here, the sugar industry stimulated the growth of the plantation of sugar cane, and from the close of the eighth century, the most fertile land was planted with sugar cane. In many cities in Sisily and Spain sugar industry was introduced soon after the Muslim arrival. Almost all Mediaeval Muslim travellers spoke of the richness of cane cultivation and sugar mills in the Muslim Empire.[17]

The making of sugar is described by contemporary twelfth century Muslim writers, such as Ibn al-Awwam, as the sugar cane being cut into small pieces and these being well trodden on or broken in a wine press, or similar presses; that its juice is placed in a clean boiler on a fire to cook and left till it becomes clear; after that it is made to boil again, until it becomes one-fourth of its first volume. It is then put into receptacles, forms, or vases made of clay of a peculiar shape, or conical. It is placed in the shade to harden, and the sugar from it is placed in the shade to aerate.[18]

The sugar industry rose to become a dominant industry in mediaeval Islam. Skills in the manufacturing of sugar seem to have remained uniquely Muslim for centuries. It involved a considerable labour force and tied up large investments in plant and machinery. The finished products gave rise to many industries, such as the making of sweets, pastries, and confectioneries. In the twelfth century onwards, the crusaders found that the Syrians were not just experts at growing sugar, but that they had also mastered the technique of crushing it under presses, extracting the juices, concentrating the substance on fire, then drying it out slowly into sugar. The Crusaders took over the industry, and followed precisely the Muslim system of production, using the same Muslim terminology.

Trade and Commerce

Arabs and Islam are marked by a kind of native impotence in economics and trade. As far back as pre-Islamic times, they organised trade caravans and spread them throughout the adjoining countries and, later, the world-wide. Actually, Islam was born in a mercantile milieu. Prophet Mohammad (pbuh) was himself a merchant and several of his companions were also merchants. So trade and commerce had their basis in Islamic philosophy, and the Muslims were the most capable traders in the world.

Trading, first and foremost, finds a good place in the Islamic religious text and the sayings of the Prophet (or Tradition). The Quran looks with favour upon commercial activity, confining itself to condemning fraudulent practices and requiring abstention from trade only during certain religious festivals. The Quran does not merely say that one must not forget one's portion of this world; it also says that it is proper to combine the practice of religion and material life, carrying on trade even during pilgrimages, and goes as far as to mention commercial profit under the name of Allah's bounty (28:77, 2:193, 4:197-8, 62:9-10).[19]

Prophet Mohammad (pbuh), himself a trader just like most of his companions, said, "The merchant who is sincere and trustworthy will (at the Judgment Day) be amongst the prophets, the just, and the martyrs; or the trustworthy merchant will sit in the shade of Allah's throne on the 'Day of Judgment', or 'Merchants are the messengers of this world and Allah's faithful trustees' on Earth."[20]

A Syrian of the late eighth century, Al-Shaibani, wrote *Kitab al-Kasb* on earning. His purpose is to show that earning a livelihood is not merely permitted, but is incumbent on Muslims. Man's primary duty is to serve Allah, but to do this properly, he must be adequately fed, housed, and clothed. This can only be achieved by working and earning. Another point made by Al-Shaybani is that money earned by commerce and industry is more pleasing to Allah's eyes than money received from the government for civil or military service.[21]

Faith and practice in Islam are closely related, and trade is even seen as an agent to spread the faith. Traders' activities flourished wherever Muslims established themselves and wherever they could reach. Muslim commerce dominated the Mediterranean till the Crusades, plying between Syria and Egypt at one end, Tunis, Sicily, Morocco, and Spain at the other, and touching Greece, Italy, and Gaul. It captured control of the Red Sea from Ethiopia, reached over the Caspian into Mongolia; and up the Volga from Astrakhan to Novgorod; Finland; Scandinavia; and Germany, where it left thousands of Muslim coins. Muslim shipping was also prominent throughout the Indian Ocean, and as early as 750, merchants from the Gulf made the long voyage to China.[22]

So there is an intricate link between faith and trade from quite early in Islam. Muslims were bound to a strict code of trade as stipulated in the Quran. The summon to record commercial deals, loans etc. is clearly stated in the religious text, which says, "Be not averse to writing down (the contract), whether it be small or great, with (record) the term thereof. That is more equitable in the sight of Allah and more sure for testimony, and the best way of avoiding doubt between you... and have witnesses when you sell one to another, and let no harm be done to scribe or witness" (Quran, 2: 282).[23]

For the accounts and making contracts, they simply had to abide by the directives of their faith, and after centuries, they perfected the practice. Islamic law and customary practice provided merchants and traders with the commercial techniques to structure and facilitate trade and exchange. At their disposal, Muslim merchants had accepted legal mechanisms for extending credit and for transferring and exchanging currencies over long distances. They issued the letter of credit, which allowed a merchant to advance or transfer a sum of money to a business associate at some distant place with the full confidence that the transfer would be expeditiously accomplished.[24]

Islam provides a legal or legislative procedure for commercial transactions. These procedures were widened to include not just Muslim inter-trading but also trading with non-Muslims. They might well have been very influential on the development of mediaeval commerce and the transmission of Muslim commercial methods to the Christian world. There is a rich Islamic literature about, and defining, economic activity, besides establishing early known economic laws. Gold currency spread widely after the Muslim arrival in the eighth century (in Spain) and after, in the form of Muslim coins and local imitations of them, which were fairly common. A subsequent Muslim innovation, the use of paper money, coming *via* China, was to revolutionise monetary transactions. Paper money was first adopted in north China by the end of the eleventh century, and the Muslims generalised the use and spread of this Chinese invention.

NOTES

1. Masood, Ehsan, *op. cit.,* p. 4.
2. S.E. Al-Djazairi (2006). *The Golden Age and Decline of Islamic Civilisation*, Manchester, p. 21.
3. *Ibid.*
4. *Ibid.*, p. 22.
5. *Ibid.*, p. 23.
6. *Ibid.*, p. 24.
7. *Ibid.*, p. 25. The use of natural power to activate machinery in paper making was a decisive breakthrough; by 950, water power was being used in fibre pounding in Bagdad. Hill dwells on how hydraulic power activates machinery and the technical processes in use. Specific natural conditions shaped particular applications of such power. Where the banks of the watercourse were high, as along the Euphrates, at Hama on the Orontes

(Syria), at Amasia on the Yeshil Irmak (further east), or at Toledo in Spain, the current of the river itself was used to turn waterwheels, which produced energy for both paper mills and other mills (flour production, for instance). The widespread use of such power is illustrated by the number of mills. In the city of Fes, in Morocco, for instance, 400 mills (operated by both water and wind power) were working in the year 1184. Not all, possibly, were geared for paper production; some such mills were used for hilling cloth, sawing timber, and processing sugar cane. Muslims compensated for the shortage of water in some parts by relying on wind power, and windmills ran mill stones.

8. *Ibid.*, p. 27.
9. *Ibid.*
10. *Ibid.*, pp. 27-28. A list of taxes on the city is given by the historian Ibn Shihna for the years 1212-3, which includes the following, among many other items of all kinds: The tax on dying factories: 80,000 Dirhems; the tax on indigo: 20,000 Dirhems; the tax on silk: 80,000 Dirhems. The daily turn-over in the cloth trade in the bazaars of the city reached 20,000 Dinars.
11. *Ibid.*, p. 29. Growing alongside such communities were other functions, such as those of financiers, warehouse owners, transporters etc. In Cairo, special quarters were reserved for cotton merchants, whilst in Bagdad, there was a special wharf for cotton merchants and a special market for the trading of cotton.
12. *Ibid.*
13. *Ibid.*, p. 32.
14. *Ibid.*, p. 33.
15. *Ibid.*
16. *Ibid.*, p. 34.
17. *Ibid.*, p. 37.
18. *Ibid.*, pp. 38-39. There is a much more detailed and richer description of the same process by the fourteenth century Egyptian, Al-Nuwairi, that the canes were firstly cut into small pieces. Then the cane is laid under the stone, which is turned over the lower stone. The juice is pressed out and flows through openings in the base to the holder, the outlet of which is stopped and the cane is then taken to another place. Here it is placed in baskets of plaited work of halfs (rushes), which are slit open below and at the sides. These are laid under the wheel of a beam, and by means of the beam, the wheel rolls on them until they have been crushed, and the rest of the juice has been pressed out. The juice that is pressed out by the stone mill and the beam (tacht) is brought to the same place. The juice is passed through a sieve, which is placed in a kind of frame. From this, it flows to a closed vessel, which is called a bastula, in a predetermined quantity. When this is filled with strained juice, it is brought to the place where it is boiled. It is then strained a second time and led to a large boiler (chabia). Into this is poured with straining all that is in the bastula.
19. *Al-Quran.*

20. M. Rodinson (1966). *Islam and Capitalism*, Paris, p. 16.
21. M.A. Cook (1970). *Studies in the Economic History of the Middle East*, Oxford, pp. 78-92.
22. W. Durant (1950). *The Age of Faith*, New York, p. 208; see also A. Pacey (1990). *Technology in World Civilization: A Thousand Year History*, Cambridge, p.12. (A Muslim tradesman said, "I want to take Persian saffron to China, where I heard it fetches a good price; then take Chinese porcelain to Greece, then Greek brocar to India, then Indian steel to Aleppo, Aleppo glass to Yemen, and Yemeni cloth to Persia." Al-Djazairi, *op. cit.*, p. 69.)
23. *Al-Quran*.
24. Al-Djazairi, *op. cit.*, p. 50.

8
Science and Technology in Modern Muslim World

Despite the fact that Islam and the Quran always speak to gain knowledge, study the universe that Allah created, and conduct research to prove it true. This proven truth is known as science, which is very essential for human civilization. Following the commandments mentioned in the Holy Quran, Muslim intellectuals gained enough scientific knowledge to last for around six centuries during the heydays of Islamic civilization.

Until about 1500, Islamic science was the most advanced in the world, and it seems beyond question that Islamic science, as transmitted to mediaeval Europe, played a critical role in preparing the ground for the Scientific Revolution. We now know, for example, that Copernicus borrowed much of the mathematics of his heliocentric system from Islamic astronomers. However, from the modern period when science gained momentum in the European world, scientific development began to decline in the Muslim world. In their competition, although much later, the Islamic intellectuals also began to give much emphasis to the religious study and learning.

Decline of Islamic Scientific Traditions

The Muslim world in the post-Mongol era gave rise to three powerful empires: the Safavi (1135-1722), the Indian Timuri (1274-1857), and the Ottoman (1343-1924). But these empires failed to restore the Muslim glory of knowledge and scientific progress, which flourished during the Abbasid and Ummayyad Spain. These empires extended over a vast area and were extremely rich in resources. They patronised the arts and sciences and vied against each other to attract the best minds of the time. In addition, several other smaller states and dynasties supported science. Yet, none of these empires was able to compete with the developments in science that were taking place after the fifteenth century in Europe. When they did realise the enormous military and economic benefits that Europeans had reaped with their science and technology, it was already too late; better-trained European armies, equipped with superior weapons, were already knocking at their doors.[1]

The Islamic scientific traditions, which flourished for eight centuries in the Muslim world, surprisingly emerged from a neutral zone of scientific inquiry. On the other hand, Christianity, which earlier rejected science eventually accepted and even supported the new scientific knowledge that emerged in Europe in the seventeenth century. By not giving much credit to Islamic Muslim scientists, European Christian scholars see Islamic scientific progress as merely a conduit through which the great works of the Greeks were safely transported to the eager minds of the European Renaissance.

The Islam which encouraged the scientific investigation and research, and as a result human created science and philosophy, but on its development science overpowered the religion and liberated the human being from superstition. The Muslim scientists were afraid of overpowering the Islamic philosophy by science and they slowly drifted away from the highest rational knowledge for which Islam had itself paved the way. Whatsoever may be the truth, it is a hard-sane fact that Islam which had earlier supported science stopped its encouragement in the sixteenth century.

The fourteenth to sixteenth centuries saw the scientific decline in the Islamic world. Although there are numerous causes for the decline of science, the overall decline of the Muslim world's intellectual, economic, social, and political conditions cannot be separated. The decline of science was actually a decline of the Muslim civilization and undoubtedly of the Islamic Golden Age. The possible causes of the decline of science in Islamic civilization may be due to (i) opposition by Islamic orthodoxy; (ii) a book written by Abu Hamid al-Ghizzali; (iii) the Mongol invasion of Bagdad in 1258; (iv) the lack of institutional support for science; (v) the disappearance of patrons; or (vi) some inherent flaw in Islam itself.

Decline of Islamic Science in the Mediaeval Period

It is critical to comprehend the facts that led to the decline of science and learning in Islamic society after it had reached its pinnacle. During the thirteenth and fourteenth centuries, Islamic science[2] went into decline; by the fifteenth century, little was left. This was accompanied by the rise of the West after the discovery of the New World, with all the wealth it brought to Europe and the eventual demise of the Silk Road, along which knowledge had flowed through various cultures.

It may be viewed that during this period, the conservative religious forces were the main hindrance to the progress of scientific observations. Besides that, debilitating warfare, economic failure, and the resulting loss of patronage, without which the sciences were unable to sustain themselves. Muslim rulers (once patrons of science) and religious scholars began to discourage scientific knowledge and began to define religion in their own way.

Crusades and Renaissance: By the early fifteenth century, Muslim civilization had experienced attacks both from the Crusaders' campaigns in Spain, Turkey, and Palestine, and from the Mongol invasion of Persia, Iraq, and Syria. The famous libraries and learning centres of the Muslim world came under catastrophic threat during these times of conflict. In Cordova, the vast majority of the city's 600,000 Islamic books were destroyed by crusading invaders. The Muslims lost countries like Sicily and Spain, which were the centres of scientific study and research. The Crusade against the Islamic world began when the Christian world informed the Europeans about the knowledge of science and the greatness of the Muslims.

It is also a well-known thesis that the Western Europeans came into contact with the Muslim scholars during the crusades against the Muslims and realised their weaknesses in terms of science and philosophy. They also developed scientific thinking and started the translation movement and renaissance. The Christian thinkers, under the influence of the Arabic progress of science and philosophy, began to study the scientific texts by translating the Greek and Arabic texts into Latin. Their translation movement that brought Islamic scientific and philosophical thought to the Latin West can be divided into three phases. The first of these three began in the late tenth century as a result of small-scale individual efforts. Although these early tenth-century translations were unplanned and unorganised but planted the seeds of what became a major intellectual tradition in the next two centuries.

The second phase of translation activity, beginning from the eleventh to the fourteenth centuries, was the most effective and successful. The Christian scholars searched for knowledge in the Muslims' past and translated thousands of manuscripts from Arabic to Latin, and they developed themselves into global fields. Thousands of those precious documents of the past, written mainly in Arabic, fill the archives of the British Library, Berlin's Staatsbibliothek, Paris' Bibliotheque Nationale, and many others. In Toledo's cathedral archive today, you can still see about 2,500 of the surviving manuscripts that scholars translated from Arabic to Latin. And it was these Latin translations that fed the scientific and philosophical revolution of the 1600s and kept the flame of knowledge alive.[3]

They called this process of development the "Renaissance". In the form of Renaissance the long cherished Muslim knowledge drifted away to the Christian world and emerged an unending power for the Europeans. From that very period they were continuously developing and never seen behind. This second phase of the translation movement was contemporaneous with the founding of several new universities in Europe. These include the universities of Bologna (1150), Paris (1200), and Oxford (1220).

The third phase of European translation activity can be classified as roughly belonging to the period between the sixteenth and seventeenth centuries. Its scope is much wider, both geographically as well as in terms of materials. However, throughout the entire translation movement, we cannot ignore the European hostility towards Islam and its prophet. This movement, on the one hand, scientifically developed the Europeans and defamed Islam and the Prophet on the other due to their unsuccessful crusader efforts. The purpose of these writings was to justify the "holy war." In these writings, Muslims appear as polytheistic, polygamous, promiscuous, worshippers of Mohammad, and wine-drinkers. These images passed from generation to generation and their remnants are still reflected in certain perceptions about Islam and Muslims held in the West.

The European Renaissance attempted to rebuild a civilization based on its antiquity. This incessant return to the past is apparent everywhere— in art, in the sciences and in poetry. At the same time, European intellectuals, writers, scientists, and artists of this period developed an intense hatred for Islam, its Prophet, and his family. During the seventeenth century, the European scientific and philosophical traditions were now able to surpass the received material from Islam. As European science surpassed Islamic scientific tradition, the propagandists of the new science began to single out Arabs as harbingers of scholasticism—mere imitators of the Greeks—whose learning was derivative and irrelevant.

Mongol Invasion: The Tartars' attack on Bagdad also led to the decline of science and philosophy in the Islamic world. The attacking Mongol armies destroyed countless manuscripts in Bagdad. The glorious period of the House of Abbas lasted until the fall of Bagdad in 1258 by Hulegu, the grandson of Genghis. For three days, the Tartar chief gave up the town to rapine and slaughter. On the third day, the thirty-seventh Abbasid Caliph Mustasin was put to death, thus ending the Abbasid dynasty. The terror, the fear, the heavy loss of material treasures and the mass murder of men of learning and wisdom stupefied and disrupted scholarship.

The destruction of culturally important parts of Islamic centres is the most important cause of the decline of the Mediaeval Islamic civilization. These destroyed centres, which were culturally important, were parts of Soviet Central Asia, Afghanistan, Iran, Iraq, Syria, Turkey, and northern Indo-Pakistan. This was accomplished by the Mongols under Chingiz Khan and Timur Lung in three waves during the thirteenth and fourteenth centuries. These devastations together started the decline and eventual fall of Islamic civilization, and the end of this bright period of classical Muslim scholarship. Massacres in some of the great cultural centres are estimated as follows:[4]

> Tirmidh, Sabziwar, Balkh, Bamiyan, Nishapur etc., entire cities; Samarqand 1000000; Khwarizm 1200000; Merv 700000-1300000; Herat 1600000; and Bagdad 1800000 in 1258. The last wave of destruction, begun by Timur Lung in 1380, continued for about a quarter century. Timur massacred 80000 people in Delhi, and in the second destruction of Bagdad (1401), he dotted the city with 120 towers of the heads of the dead. Libraries, colleges, hospitals, and mosques (which were also institutions of learning) were burnt and destroyed. Irrigated lands were converted to deserts occupied only by nomadic herdsmen and marauders.

Non-Cooperation of Muslim Rulers: In Western Africa, the triumph of Patristicism under the third Almohade sovereign, Ya'kub al-Mansur, and the uprise of Berber fanaticism turned back the tide of progress. They arrested the civilization of centuries and converted the seats of learning and the arts into centres of bigotry and ignorance. The settlement of the Corsairs on the Barbary Coast and the anarchy which prevailed in Egypt under the later Mamluks discouraged the cultivation of peaceful knowledge. In Asia, the decadence of the Timurid dynasty, the eruption of the wild and fanatical Uzbegs, and the establishment of their power in the capital of Timur, destroyed the intellectual vitality of the people.[5]

Unlike his forefathers, Abbasid ruler Al-Mutawakkil, neglected the study of nature. The spirit of scientific investigation was replaced by credulity and superstition. Those who loved free inquiry and advocated mental freedom found themselves ostracized. The slightest deviation from fixed prejudices was punished with a severity. Philosophy was under a cloud, and many brilliant men, obstructed in their path, took refuge in mysticism. The colleges and universities were closed; literature, science, and philosophy were interdicted; and the Rationalists were hunted from Bagdad.[6]

Religious Conservation: While European scholars were studying science and nature, Muslim scholars were studying religion and theology and had completely abandoned their scientific interests. Religious scholars believed that the introduction of foreign science into the Islamic way of life was risky because it completely rejects mindless superstitions. As a result, no more borrowings took place from the world outside Islam. Hellenistic philosophy now came to be seen as a danger to the faith. The Muslim scholars were afraid of the development of modern scientific developments as it became a general belief that without science, humankind was helpless before wind and storm, ravaged by plague and disease. After the development of science and technology, it has become a global need for survival. It has made itself free from religious bonds, individual opinions and beliefs.

Secular Pursuit of Science: At present, science has gained a separate and most important global field of knowledge, based on purely observation and experimental research, free from any religious bond. As it was promoted by scientists of various faiths, science has now become a secular pursuit, and it is impossible for it to be otherwise. The secular character of science does not mean that it necessarily repudiates the existence of the Divine. The modern scientists believe in a Allah who created and ordered the universe, but they do not believe in a Allah who intervenes at will to change the course of a planet, to postpone an eclipse, to alter weather patterns against the dictates of fluid physics, or to change the rules of the cosmic game in any other way. Scientists are free to be as religious as they please, but science recognises no law outside its own.

Undoubtedly, the old theories of scientists are the base of knowledge but the scientific developments and research from those times to the modern period have gained much advancement, and the researcher should consult the latest observations. For example, a student of physics who studied optics from Ibn al-Haytham's *Kitab-al-Manazir* or mechanics from Newton's Principia would be infinitely better advised to study a modern book which includes the cumulative work of the thousands of workers who, since these great men, have assiduously worked to improve, expand, generalize, and simplify the subject.

It is really a matter of concern that the Muslims in modern times, who comprise one-fifth of all humanity in the entire world, have drifted away from the knowledge of science. But there is hope for the future as growing number of Muslims come to understand the need for attitudinal changes if science is to thrive once more in Muslim lands. The Islamic powers are leaning towards military strength, political power, and economic prosperity, which are essential credentials for acquiring knowledge and scientific development.

While Europeans were continuously developing scientific approaches right from the Renaissance period and emerged as the most hi-tech nations with scientific advancement, economic prosperity, and unbeatable military power, the Islamic nations were traditional, agrarian, and even mediaeval in ideology and progress. Now, there is no exaggeration to say that the modern science has become the sole property of the West. It is really a matter of shame that Muslim countries after the great World Wars never learnt the lesson and remained busy in the pleasure and amusement. They never tried to educate their people and advance themselves scientifically and militarily. They remained mere followers of the West and spectators of the scientific propaganda of the West.

Scientific Position of Muslims in Modern Period

The scientific progress and technological inventions in Europe, such as the steam engine, the hydraulic press, technologies used in the extraction of coal and the purification of metals, and military technologies, would be looked upon by Muslims as "wonders of European science". This whole process of change, which had been taking place in Europe for over two centuries, was brought home to Muslims when Napoleon arrived in Egypt in 1798. His army had superior weapons, better training, and it was accompanied by scientists specialising in diverse branches of science and several new technologies. Muslims ascribed their military subjugation to a lack of science and technology, and their leaders told them that their lost glory would be restored as soon as they caught up with Europe in science and technology. Religious scholars used their influence to support this call for the acquisition of science.[7]

In modern times, the Islamic world has been disintegrated into numerous independent nation states, many of which have been colonised by European forces or have gone to war with one another, even after gaining independence. During the eighteenth and nineteenth centuries, the entire Middle East, a large part of Africa, the whole middle belt of Asia, and the Malayan archipelago were under Muslim control; but by the end of that fateful century, a large part of this territory had come under Russian, British, French, Portuguese, and Dutch influence or direct control; by the middle of the nineteenth century, there was nothing left of the power, might, and splendour of the old Muslim world. Neither the peoples of these states nor their colonial rulers had prioritised education or socio-economic reform, leaving these countries to struggle for identity and survival in a fast-paced modern world. All these territorial collapses were due to the lack of science and technology among Muslim countries.

People were taught European scientists rather than Ibn Sina, Al-Biruni, Ibn al-Haytham Al-Khwarizmi, and Ibn al-Shatir during colonial education policy. Their tradition was violently plucked out of Muslim lands, leaving them bereft of historical depth. Muslim societies have become victims of a cultural schizophrenia in which the past appears as a ghost to be exorcised. More ridiculously, even the Muslim theologians also did not heed towards their past and boosted the Muslim community only with religious instruction and began to prepare them not for worldly affairs but for life after death.

Although Western scientists were highly advanced, Muslims were completely reliant on Western science and discoveries. They even imported military equipment and weapons from them. However, the failure of Arabic science in the mediaeval period gave birth to modern science, but Muslims of the modern world never benefited themselves from their historical past. The Muslims simply assert that modern science is a refined version of "our own science," which Europe developed and returned to.

The Muslims of the modern period never followed their forefathers and did not try to develop their warfare technology and strategy. World Wars I and II were very decisive and finally, proved that Western powers were the best in terms of science and military technology. The hi-tech war waged by the West against Iraq and Israel against Palestine were vivid illustrations of this ideology. The civilization of Islam has paid a heavy price for having failed to acquire science and technology.

Rejuvenation of Scientific Traditions in Modern Muslim World

The Ottomans were possibly the first Muslims to recognise their lack of technological knowledge. This came about through military defeats. The peace treaty signed on January 26, 1699, at Carlowitz was the end of a long chapter in their history and the entire equation between them and Europe had changed. In 1718, they were compelled to sign the treaty of Passarovitz by losing Belgrade. By 1774, the Ottomans had lost control of the Black Sea to the Russians. The Ottomans were now convinced of the need for reform. As a result of these setbacks, Caliph Salim III undertook far-reaching military and political reforms in 1792. The acquisition of science and technology were the main features of this reform.[8]

Almost all intellectuals and reformers began with a perception that Islam is a religion that supports the acquisition of knowledge: modern science is knowledge; the acquisition of knowledge is an obligation of all believers; Muslims must, therefore, acquire science. This knowledge, it is further argued, cannot contradict Islam, for science studies the work of Allah and the Quran is the word of Allah, and there can be no contradiction between the two. However, this rallying cry to acquire scientific knowledge was ultimately unsuccessful because most of the champions of this rallying cry were not scientists or religious scholars; rather, they were reformers who saw modern science as a means to power and progress. They saw Muslims in need of both, and hence they used Islam to justify their agenda.

If one takes the debate started by Ernest Renan in Paris in 1883 as a starting point, then Islamic perspectives on modern science began to emerge in the closing years of the nineteenth century. Muslim scholars, such as Jamaluddin Afghani, very clearly responded to the thesis of Renan, and an attempt to acquire knowledge began to be seriously planned by the Muslim scholars. In the new discourse on Islam and science during these periods, books started to appear with verses from the Quran purporting to confirm scientific theories. The incomplete *tafsir* (commentary) of the Quran by Sir Syed Ahmad Khan, published in 1879-1898 and another later scientific *tafsir*, *Tafsir al-Ilmi,* of Abu Hajar and Ahmad Umar, are the most important in this regard.

Egypt

After developing themselves scientifically, the Europeans began the policy of colonialism, and in this process, they entered India and the rest of the world, with scientifically advanced military systems to conquer the territories for their economic benefit. Sometimes, the European conquest directly or indirectly benefits the Muslim countries. The arrival of Napoleon in Egypt was one of the examples.

Mohammad Ali (1769-1849): After Napoleon's army was driven out of Egypt and Mohammad Ali took control, one of Ali's most important goals was to modernise Egypt by acquiring modern science. He set up new schools and training colleges where modern science was taught by foreigners hired for inflated salaries. Between 1825 and 1836, his government founded military and naval institutions for training officers and soldiers in the various military professions on modern lines, including some disciplines of modern science. But these measures produced no science. It was like attempting to have a body without a backbone. All that Mohammad Ali could do was rupture old traditions without succeeding in establishing a new system.[9]

Mohammad Ibn Ahmad al-Iskandarani: Another work of much greater impact was by an Egyptian physician, Mohammad Ibn Ahmad al-Iskandarani, and was published in 1880. His book has a long title: *The Unveiling of the Luminous Secrets of the Quran in which are Discussed Celestial Bodies, the Earth, Animals, Plants, and Minerals*. Three years after the publication of this work, Al-Iskandarani published another book in 1883 on the Divine Secrets in the World of Vegetation and Minerals and in the Characteristics of Animals. In both works, Al-Iskandarani explained verses of the Quran to prove the presence of specific scientific inventions in the Quran. This trend of writing a scientific commentary on the Quran became so popular that general surveys on the Quranic exegeses had to invent a new category of *tafsir* literature to accommodate this genre. After the trend was established by Al-Iskandarani, several other works of this kind appeared in the Arab world.[10]

Mohammad Abduh (1849-1905): The Quranic *tafsir* based on scientific interpretation became a common feature of Muslim works in the early part of the twentieth century. One of the most influential was that of Al-Afghani's Egyptian student, Mohammad Abduh. His disciple Rashid Rida compiled a series of lectures on Quran given by Abduh in Cairo and named as *Tafsir al-Manar*. *Tafsir al-Manar* was finally published in 12 volumes in 1927. This was one of the most influential works during the first half of the twentieth century, as can be judged from the number of editions of this work that appeared between 1927 and 1950.[11]

Tantawi Jawhari (1870-1940): The scientific exegesis of the Quran reached a high point in 1931, when a twenty-six-volume *tafsir* was published by Egyptian scholar Tantawi Jawhari. His *tafsir*, called *Al-Jawahir fi Tafsir al-Quran al-Karim* (Pearls from the Tafsir of the Noble Quran), appeared with illustrations, drawings, photographs, and tables. In his introduction to the work, Tantawi says that he prayed to God to enable him to interpret the Quran in a manner that would include all sciences attained by humans so that Muslims could understand the cosmic sciences. He was convinced that Quranic chapters supplemented what modern science was discovering.[12]

Turkey

Damad Ibrahim Pasha (1718-1730): In Ottoman Turkey, the Grand Vazir Damad Ibrahim Pasha introduced a number of measures aimed at the modernization of Turkey, including naval reforms and the use of the printing press. Scientific geography became part of the curriculum of military schools. New training centres and schools were introduced on

European models, often with the help of European experts and converts. There were some visible results. A two-arc quadrant was invented by Mehmed Said, the son of the Mufti of Anatolia, for use by artillerymen; treatises on trigonometry, new works on medicine, as well as translations of certain European scientific and philosophical writings, started to appear. New naval schools of mathematics established in 1773, produced a new corps of engineers and artillery.[13]

The nineteenth century witnessed a series of reforms, all aimed at modernising Turkey. New schools were opened. European languages were taught; scientific texts were translated into Turkish; the first modern census and survey were carried out in 1831. In 1845, a commission of seven eminent men was put in charge of proposing educational reforms. Its report, submitted in August 1846, called for the establishment of an Ottoman State University, a network of primary and secondary schools, and a permanent Council of Public Instruction. However, these reforms were ineffective, and critical thinkers openly criticised their own rulers.[14]

Mustafa Kemal Pasha (1881-1938): During World War I, the Sultan had lost much public support, which led to the emergence of Mustafa Kemal Pasha, later known as Kemal Ataturk (1881-1938), who launched a war of liberation to save Turkey from falling into the hands of European powers. On March 3, 1924, he abolished the Caliphate, banished all members of the Ottoman Sultanate from the Turkish territory, and Turkey became a secular republic. Everybody knows what he did to Islam in Turkey, but he had a certain intuitions, and certain visions of things.

The first thing that Kemal Ataturk did was to say that in order for Turkey to stand on its feet as a modern "secular" state, what it had to do was to learn about the history of Western science. Ataturk sent the first student to study the history of science anywhere in America to Harvard. The first person to enter the Ph.D. programme in the history of science at Harvard University is a Turk, Aideen Saeeli.[15]

For Mustafa Kemal and his associates, civilization meant European civilization. For them, everything related to Islam meant backwardness. They banned traditional dress. Mustafa Kemal declared that his legacy was not Islam but science and reason. Islam was now perceived as the greatest enemy of science, which was seen as the only means of progress and civilization.[16]

Badiuzzaman Said Nursi (1877-1960): Said Nursi opposed the secular ideas of Mustafa Kemal Ataturk. For that, he was exiled to western Anatolia in 1925 for twenty-five years, along with thousands of other Muslims. His writings during exile were published as *Risale-i Nur*.

Said Nursi had considerable knowledge of modern science, especially of physics. He attempted to show that there could be no dissonance between the Quran and the modern physical sciences. He found modern science to be useful in conveying the message of the Quran. Nursi's profound spiritual insights led many of his countrymen and other Muslims back to their religion against a strong state-sponsored secularization that had all but erased Islam from the Turkish public space.[17]

India

In India, the disintegration of the Mughal Empire led to its colonization. The fatal battles of Plassey (1757) and Buxar (1764) consolidated the British hold over Bengal and paved the way for further conquests in Bihar and Orissa. When the British won the battle against Tipu Sultan, the visionary ruler of Mysore, in the closing years of the eighteenth century, the fate of India was sealed, for Tipu's army was the last real resistance against the colonisation of India. Ultimately, India became a British colony and remained enslaved for about two hundred years.

Sir Syed Ahmad Khan (1817-1898): The Indian Muslims, who were scientifically far behind like the Muslims of the rest of the world, began to follow the Europeans to learn science and English literature. But it was a really very tough job as most of the Ulema classes were religiously very conservative. Under British rule, some Indian social reformers began to propagate the learning of science and proved it through Quranic interpretations. Among the early leaders of this approach to Islam and science was Syed Ahmad Khan, who attempted to harmonize Islam and modern science by reinterpreting the Quran.

It was Sir Syed Ahmad Khan who first attempted to make a transition from mediaeval to modern Islam. He advocated the study of science along with religious studies. He tried to build a bridge between Islam and modernity. Sir Syed Ahmed Khan wrote, "…if the Muslims be a true warrior and thinks his religion is right, then let him come fearlessly to the battleground and do unto Western knowledge and modern research what his forefathers did to Greek philosophy. Then only shall the religious books be of any real use. Mere parroting will not do."[18]

Syed Ahmad's efforts were directed towards educational reforms. He concluded that Muslims were backward because they lacked modern education. He equated modern education with modern science. He established a Scientific Society, convening its first meeting on January 9, 1864. In 1867, Syed Ahmad Khan and the Scientific Society moved to Aligarh, where he procured a piece of land from the British government to establish an experimental farm. He devoted all his energies and a portion of his personal income to the Scientific Society he established for the promotion of science. The Scientific Society had four specific goals:[19]

1. To translate into such languages as may be in common use among the people those works on arts and sciences that, being in English or other European languages, are not intelligible to the natives;
2. To search for and publish rare and valuable oriental works (no religious work will come under the notice of the Society);
3. To publish, when the Society thinks it desirable, any (periodical) which may be calculated to improve the native mind; and
4. To have delivered in their meetings lectures on scientific or other useful subjects, illustrated, when possible, by scientific instruments.

Sir Syed Ahmad Khan was an ardent believer in the utility of modern science, a devout Muslim, an educator, and a man with a cause. He dedicated his life to the uplifting of Muslims in India. He wanted them to acquire modern science and be among the honourable nations of the world. On May 10, 1866, he established another organization—The Aligarh British Indian Association—to promote scientific education. For that, Sir Syed Ahmad Khan went to England with his two sons, Syed Hamid and Syed Mahmud. They left India on the first of April, 1869; to pay for this trip, Syed Ahmad Khan had to mortgage his ancestral house in Delhi and borrow 10,000 rupees. While in England, Khan was awarded the title of "Companion of the Star of India" by the Queen. His stay in England convinced him of the superiority of the British. He returned home on October 2, 1870.[20]

During his stay in England, he visited the Universities of Oxford and Cambridge and a few private schools, including Eaton and Harrow; these would later serve as models for his own "Mohammadan Anglo Oriental College", established seven years after his return to India. Initially, it was a school "Madarsatul Uloom Musalmanane Hind" founded in 1875 in Aligarh for study of science to the youth of India. He was nominated as a member of the Vice Regal Legislative Council in 1878; ten years later, he was knighted as Knight Commander of the Star of India. In 1889, he was awarded an honorary degree from the University of Edinburgh.[21]

In 1886, Sir Syed Ahmad Khan started another institution, "The Mohammadan Educational Conference," which organised various conferences in major cities for several years. Personally, he was convinced that Islam and modern science were perfectly aligned. To prove his views,

Khan decided to write a new commentary on the Quran. This was a feverish effort that began in 1879. When he died in 1898, this most important work of his life was still incomplete, yet it was published in incomplete form during the year 1879-1898. His unfinished commentary attempted to rationalise all aspects of the Quran.[22]

Many years after his death, his Mohammadan Anglo Oriental College was elevated to a university named Aligarh Muslim University in 1920, whose students continuously took part in scientific achievements and nation-building. He was not only a thinker but also a practical man who set up institutions that influenced, and continue to influence, the course of education, intellectual thought, and discourse on Islam and science.

Syed Ameer Ali (1849-1924): Syed Ameer Ali was another great Muslim visionary of Islam and science. He wrote a book, "*The Spirit of Islam*" and proved that true Islam is revolutionary, rational, and progress-oriented. His views on Islam and science can be summarised as follows:[23]

(i) The Holy Qur'an and the sayings of the Prophet give supreme value to knowledge. Knowledge is to be understood as including science. This is what motivated the early Muslims to study science.

(ii) Aristotelian philosophy and rationalist thinking were entirely in accordance with Islam. The Muslim philosophers and scholars, Al-Kindi, Al-Farabi, Ibn Sina, Ibn al-Haytham, Ibn Rushd, are true heroes of Islam.

(iii) Syed Ameer Ali identifies those most responsible as Al-Ashari, Ibn Hanbal, Al-Ghizzali, and Ibn Taymiyya.

(iv) Science needs to be brought back from the West into Islam; it is not something foreign to Islam and not by any means un-Islamic. In his opinion, Islam needed to be rescued from the *Mujaddids* and *Imams*, and the minds of the Muslims were freed from the bondage of literal interpretation.

Syed Jamaluddin al-Afghani (1838-1897): Syed Jamaluddin Afghani was another pro-modernity and pro-science. Jamaluddin al-Afghani migrated to Egypt and called him Al-Afghani. The famous reformer, a rather maverick figure of the nineteenth century, was at once a philosopher, political figure, Pan-Islamist and anti-Caliphate organizer. He was certainly a very influential person in the nineteenth century, and was responsible, directly and indirectly, through his student Mohammed Abduh, for the so-called reforms that took place in the 1880s and 1890s. Afghani was a promoter of science and called science friendly to the Islamic spirit.

Jamaluddin was interested in Western science, though he had very little knowledge of it, and he was also very much interested in the revival of the Islamic world. Firstly, he came up with the view that science *per se* is what has made the West powerful and great. And the West is dominating the Islamic world because it has this power in its pocket. And since this is being allowed, since this is being done, there must be something very positive about this science, that science itself is good, because it gives power. Secondly, he argued, science came from the Islamic world originally, and therefore, Islamic science is really responsible for the West's possession of science and the West's domination of the Islamic world itself. And therefore, all the Muslims have to do is to reclaim this science for themselves in order to reach the glories of their past and become a powerful and great civilization.[24]

Unlike Sir Syed Ahmad Khan, Afghani did not make a serious attempt to reinterpret Muslim theology. Afghani was firstly a great advocate of Islam and bitterly criticised Sir Syed Ahmad and accused him repeatedly of heresy and deviating from Islam. He wrote his Persian refutation of Syed Ahmad Khan's Nechari (naturalistic) ideas, "The Truth of the Nechari Religion and an Explanation of the Necharis (1881)."[25] Afghani's also inspired the Muslim masses to resist the yoke of foreign domination and infused in them a sense of purpose and pride.

In his 1882 lecture in Calcutta, he said, "Thus I say: If someone looks deeply into the question, he will see that science rules the world. There was, is, and will be no ruler in the world but science .... The benefits of science are immeasurably great, and these finite thoughts cannot encompass what is infinite.... thanks to the Islamic religion, a philosophic spirit arose among them.... This was why they acquired in a short time all the sciences, with particular subjects that they translated from the Syriac, Persian, and Greek into the Arabic language." While attacking the Indian Ulema, he said, ".... They never ask the causes of electricity, steamboats, and railroads .... our Ulema at this time are like a very narrow wick on top of which is a very small flame that neither lights its surroundings nor gives light to others ...".[26]

While in Paris, he wrote the famous "*Answer to Renan*". Renan (grandfather of rationalism in nineteenth century French philosophy) built a thesis on "Islam against science" on the basis of the Orientalist studies of the previous two centuries. Renan's thesis gave birth to Goldziher's (Hungarian scholar of Islam) influential doctrine that posited a supposed Islamic Orthodoxy against foreign sciences. Renan's main point was that early Islam and the Arabs who professed it were hostile to the scientific and philosophic spirit, and that science and philosophy had entered the Islamic world only from non-Arab sources. In his response, Jamaluddin Afghani sought to defend Islam by broadening the arguments and blamed all religions for being intolerant and being an obstacle to the development of science and philosophy. With time, he said, all people, including Muslims, should learn to overcome these obstacles.[27]

Afghani says that science should not be known as Arabic science, European science, and Muslim science or Christian science. Men must be related to science, not science to men. The father and mother of science are proofs, and the proof is neither Aristotle nor Galileo. According to Afghani, the man most accused of destroying science in Islam was Abu Hamid al-Ghizzali. Afghani said that Islamic religion is the closest of religions to science and knowledge, and there is no incompatibility between science and knowledge and the foundation of Islamic faith.

NOTES

1. Iqbal, Muzaffar, *op. cit.*, p. 154.
2. Islamic science is actually knowledge that is directly or indirectly mentioned or indicated in the Holy Quran and instructed by the Prophet. But here it really means a science that developed during the heydays of Islam or during the reign of the Muslim rulers. It is a science that takes the natural world as a sign of the One who created it in the first place and Who is continuously, singularly, and uniquely its Sustainer. Furthermore, it is a science that is practiced from within the traditional Islamic view of the natural world in the organic and dynamic relationship between scientific investigation and the real and true needs of humanity.
3. Ameer Ali, *op. cit.*, p. 400.
4. S. Waqar Ahmed Husaini (1980). *Islamic Environmental Systems Engineering*, London. p. 19. (Some of the Abbasid relatives took refuge in Egypt. The Mamluk Sultans searched for the Abbasid relatives who had taken refuge in Cairo, and recognised one of them as a Caliph. Now the Abbasid Caliphs were only titular and their functions remained purely spiritual. Ameer Ali, *op. cit.*, p. 402).
5. Ameer Ali, *op. cit.*, p. 402.
6. Al-Hassani, *op. cit.*, p. 22.
7. Bhargava, *op. cit.*, p. 186.
8. Iqbal, Muzaffar, *op. cit.*, p. 155.
9. *Ibid.*, pp. 155-56.
10. *Ibid.*, p. 158.
11. *Ibid.*, p. 179.
12. *Ibid.*, pp. 179-80.
13. *Ibid.*, p. 180.

14. *Ibid.*, pp. 160. It was surprising that in the fifteenth century, certain Jewish refugees, fleeing Spain, had set up printing presses in Constantinople in 1493 or 1494; this was followed by similar presses in other cities. But Arabic and Turkish texts could not be printed due to a ban that was not lifted until July 5, 1727.
15. *Islam and Modern Science* by Seyyid Hossein Nasr, https://www.academia.edu/2521834/Islam_and_Modern_Science_1982.
16. Iqbal, Muzaffar, *op. cit.*, p. 161.
17. *Ibid.*, pp. 161-62.
18. *Ibid.*, p. 178.
19. Bhargava, *op. cit.*, p. 56. Sir Syed Ahmad Khan directly or indirectly tried to comment on great events of Islamic history, such as the fall of Adam and Eve, the Great flood, *Meraaj* incident as mere symbol to teach the morality to the people, and results of the sin and evil, and closeness of the Prophet to Allah. He told that Quran is the book of morality and not a book of science. However, over his efforts of modernity and science, and comments Islamic jurisprudence, he faced numerous *Fatwas* of *Ilhad* (apostasy) and *Kufr* (unbelief) issued by the Ulema. The *Mutawalli* (keeper) of the Holy Kaaba declared him to be an enemy of Islam and *Wajib-i-Qatl* (deserver of death).
20. Iqbal, Muzaffar, *op. cit.*, pp. 167-68.
21. *Ibid.*, p. 168.
22. *Ibid.*, pp. 169-70. His work was severely criticised by religious scholars and other Muslim intellectuals, who pointed out his lack of training in Islamic sciences and his inability to use Arabic sources; his zeal to show the agreement between the Word and the Work of God earned him the pejorative title of "Nechari" (naturalist).
23. Bhargava, *op. cit.*, p. 58.
24. Iqbal, Muzaffar, *op. cit.*, p. 172. Afghani believes that acquisition of this science of the West by the Muslims is, no more no less, than the Muslims claiming their own property which has somehow been taken over by another continent and the Muslims just want back what is really their own.
25. Bhargava, *op. cit.*, p. 60.
26. Iqbal, Muzaffar, *op. cit.*, p. 174.
27. *Ibid.*, pp. 175-76.

9
Post-Colonial Islamic World
Scientific Development and Devastation

The colonial world, which included most of the countries of the world, and undoubtedly the Muslim world was also among them, left the Muslims unaware of scientific teaching and research. Their knowledge of modern science was limited; their education was limited; books were scarce; there were no scientific laboratories; and the great tradition of scientific research had all but vanished from their lands. In addition, the whole Muslim world was in a state of internal strife. But when they got freedom from European slavery, some of them grew rapidly. For the Muslim world, the post-1950 era has been like a rude awakening from a mediaeval siesta. The Muslims faced challenges which they had never faced in their long history. Most of these new challenges are somehow related to science and technologies produced in the West.

All the new Muslim states that emerged from colonial bondage embarked upon rapid modernization plans. They sent their best minds to the West to acquire science, they imported technologies, and they attempted to look like their former colonial masters in nearly all respects, from governance to social norms. Science and technology became the most coveted commodities in these new states. Almost all of these states established ministries of science and technology, developed official policies to improve science education, and diverted significant sums from their budgets to establish new scientific research institutions.[1]

The Muslim nation-states sent thousands of students to Europe and North America to obtain doctorates in various branches of science with the hope that they could jump-start the production of science upon their return. The results could have been expected: armed with their Ph.Ds., these students returned home to find there was no infrastructure to do science. Libraries with the latest journals, laboratories with working instruments, industry in need of scientific and technological solutions to its problems, large pharmaceutical and defence industries—none of these features of modern scientific research were present in the Muslim world.

Despite its educated population, the Muslim world was completely reliant on the West for scientific and technological assistance. Notwithstanding, some of them emerged as powerful scientifically and, to some extent, developed their own science and technology. However, they faced a serious threat from the West as the Western powers imposed a ban on scientifically developing Muslim nations and tried to finish their power by war or treachery. In this course of emergency, some Muslim countries were mere spectators, and no process was initiated to unite them against the West. It resulted in the formal and informal military victories of the West over Muslim countries such as Iraq, Afghanistan, Egypt, Syria, Libya, Palestine etc. Israel, a very tiny state of merely 90 lakh people but scientifically and militarily strong, has challenged 57 Muslim prosperous countries by bombarding the Gaza Strip, occupying Palestinian territories and one of the most holy Muslim places, Masjid-al-Aqsa and Bait-al-Muqaddas.

Israel's attack on Gaza and Palestine and the military retaliation by Hamas of Palestine shows a look over the Muslims' interest in science and military technology and courage to fight the enemy. Fortunately, Iran and Turkey emerged as powerful Muslim countries after the fall of Saddam Husain of Iraq, and it seems very likely that they were trying their best to acquire knowledge and scientific research and provide technology to the other needy Muslim states. Palestinians and a small group of Hamas' resistance indicate the same.

Lack of Unity and Efforts among Muslim Nations

The major reason for the scientific drawback of the Muslim countries is the lack of unity, which partially destroyed the powerful Muslim countries such as Iraq, Afghanistan, Egypt, Syria, Libya, Palestine etc. The Muslim countries are prosperous oil producers and suppliers to the entire world, but they are practically absent in the field of scientific progress. It is strange, for example, that the Organization of Islamic Conference, with its headquarters in Jeddah, has a permanent "Committee on Scientific and Technological Cooperation" with the expressed aim of promoting science and technology in the Muslim world, yet it has no scientific institutions, laboratories, or journals. Almost all Muslim states have ministries and ministers of science and technology, who ceaselessly issue statements on the need to acquire modern science, but none of these fifty-seven

Muslim states produces any science worth its name, and most able Muslim scientists live outside these states.[2]

The Muslim countries are jealous of each other and hardly sympathetic to the Muslims of the rest of the world. But, besides the disunity among the Muslim states, Iran and Turkey are progressing scientifically and militarily and have undertaken the impossible to prepare the rest of the Muslim world for a united front against the West. It has been recently observed that the Muslim community as a whole has seriously inclined towards education and the study of science, and they hope that scientific advancement is the only path of survival in the future world.

Scientific Interpretation of Holy Quran

In the post-1950 era, a new Islamic perspective on modern science emerged, which was more modern than the descriptions of late nineteenth century Muslim scholars. Many Muslim scholars began to interpret the Holy Quran scientifically and proved the nexus between science (or modern science) and Islam. These new issues have arisen as part of a greater process of change in the Muslim world and a brief overview of these processes, which have ushered the Muslim world into the twenty-first century.

The Muslim scholars boosted the Muslim nations' scientific advancement. Many Muslim nations have also begun to propagate Islam as a justifier for science. It has become a notion that in the new system of education in the modern Muslim world, a new course on theology is not needed; instead, all courses should be changed into courses of theology. Some Muslim nations, such as Pakistan, introduced a new curriculum of science in basic schooling to mingle science with theology or Islamic study.

The zeal of advancement gave new impetus to the process of writing *Tafsir* (commentary) literature—the *Tafsir al-ilmi*, the scientific Tafsir of the Quran—which attempts to prove scientific facts and theories from the verses of the Quran. Such works include *Mujizat al-Quran fi Wasf al-Kainat* (The Miracles of the Quran in the Cosmos) by Hanafi Ahmad (1954); *Al-Islam wa tihh al-Haditha* (Islam and Modern Medicine) by Ismail Abd al-Aziz; *Al-Nazariyya al-Ilmiyya fil-Quran* (Scientific Theories in the Quran) by Matb al-Itimad (1942); *Creation of the Heavens and the Earth in Six Days in Science and in the Quran* by Hasan Atiyyah (1992); and many others.[3]

Maurice Bucaille (1920-1998): The scientific interpretations of the Holy Quran gained much momentum when a French physician, Maurice Bucaille, published his enormously popular book, *La Bible, le Coran et la* *science : Les ecritures saintes examinees a la lumiere des connaissances modernes* (The Bible, the Quran, and Science: The Holy Scriptures Examined in the Light of Modern Knowledge) in 1976. Bucaille's book has been translated into every language spoken in the Muslim world. Bucaille, who was the family physician of the Saudi King Faisal, attempted to show that the Quran contains scientifically correct information about the creation of the heavens and earth, human reproduction, and certain other aspects of the natural world, whereas the Bible does not.[4] Bucaille's work is a forerunner to numerous other works that attempt to interpret the Quran on the basis of modern scientific knowledge. In all such works, the Quranic vocabulary is placed within the framework of modern science and its verses are interpreted to show the existence of scientifically correct knowledge in the Quran.

Keith Moore (1925-2019): A Canadian embryologist, Keith Moore, was another to write a scientific interpretation of the Quran. He wrote, "I was astonished by the accuracy of the statements that were recorded in the seventh century, before the science of embryology was established. Although I was aware of the glorious history of Muslim scientists in the tenth century and of some of their contributions to medicine, I knew nothing about the religious facts and beliefs contained in the Quran and Sunnah. It is important for Islamic and other students to understand the meaning of these Quranic statements about human development, based on current scientific knowledge."[5]

During the Seventh Medical Conference held by the Commission at Dammam, Saudi Arabia, in 1981, Moore said that "it has been a great pleasure for me to help clarify statements in the Quran about human development. It is clear to me that these statements must have come to Mohammad from God, because almost all of this knowledge was not discovered until many centuries later. This proves to me that Mohammad must have been a messenger of God." During the question session, when Moore was asked, "Does this mean that you believe that the Quran is the word of God?" he replied, "I find no difficulty in accepting this."[6]

Abdul Majeed al-Zindani (Born in 1942): In the process of scientific interpretation of the Holy Quran, a Commission for Scientific Signs of the Quran and Sunnah was established in Saudi Arabia by Abdul Majeed al-Zindani of Yemen with the backing of the World Muslim League in 1984 (1406 AH). It was an organisation established to publicise what it calls scientific signs found in the Quran and Sunnah. It has the following six goals and objectives:[7]

(i) To lay down governing rules and methods (for studying) scientific signs in the Holy Quran and Sunnah;

(ii) To train a leading group of scientists and scholars to consider the scientific phenomena and the cosmic facts in the light of the Holy Quran and Sunnah;

(iii) To give an Islamic Character to the physical sciences through introducing the conclusion of approved researches into the curricula of the various stages of education;

(iv) To explain, without constraint, the accurate meanings of the Quranic verses and the Prophet's Traditions relating to Cosmic Sciences, in the light of modern scientific finds, linguistic analysis and purpose of Shariah;

(v) To provide Muslim missionaries and mass-media with Dawah;

(vi) To publicise the accepted research in simplified forms to suit the various academic levels and to translate those papers into the languages of the Muslim world and the other living languages.

The Commission has also started publishing several books dealing with the scientific miracles of the Quran in various fields such as embryology, botany, geology, astronomy, and cosmology. It organised many international conferences in various countries, which hosted splendid ceremonies where Western scientists were invited to receive attention and patronage from princes and other high officials of kingdoms and states. These conferences have covered all the verses of the Quran that have any relevance to various branches of science such as embryology, geology, and medicine. Similar state-sponsored programmes were initiated in Pakistan, Jordan, and other Muslim countries. The audio-visual recordings of these conferences are available on scores of websites, and numerous books have been published in various languages that use material from these conferences.[8]

Seyyed Hossein Nasr (born in 1933): The most representative voice of the traditionalist discourse on modern science is that of Seyyed Hossein Nasr of Iran. Beginning with *An Introduction to Islamic Cosmological Doctrines* (1964), Nasr's scholarly writings have explored various aspects of Islam's relationship with science over the last forty-two years. These works are part of a corpus of writings that addresses almost all aspects of Islam and its civilization—from the Islamic concept of Ultimate Reality to sacred art and architecture. His works on science have explored the essential features of traditional sciences of nature as well as issues concerning modern science. His training in the history and philosophy of modern science, and the inner resources gathered from traditional wisdom, are a unique combination, making him the chief expositor of a clear and insightful Islamic perspective on science.[9]

Syed Mohammad al-Naquib bin Ali al-Attas (born in 1931): He is a Malaysian Muslim philosopher. Syed Naquib al-Attas made significant contributions to the contemporary world of Islam in the domains of the Islamization of contemporary knowledge and of Muslim education. He formulated this at the First World Conference on Muslim Education, held in Makkah (1979), and founded and directed the International Institute of Islamic Thought and Civilization (ISTAC). ISTAC aims to create the perfect man with the capacity and personality in accordance with the teachings of Islam. He is among the few contemporary scholars who are thoroughly rooted in the traditional Islamic sciences, being a competent master in theology, philosophy, metaphysics, history, and literature. The objectives of the ISTAC are as follows:[10]

1. Formulating, researching, defining, and outlining the basic concepts of Islam relating to the issues of culture, education, science, manner, and style as well as the sources of knowledge faced by Muslims today;
2. Resulting and providing the answers based on Islam on many challenges of intellectual and modern culture and various schools of thought, religion and ideology;
3. Assessing the meaning and philosophy of art and art of Islamic development and giving form and guidance towards the Islamizing field of literature and education of the arts;
4. Investigating, reviewing, and writing *Tamaddun* in Malay;
5. Formulating Islamic education philosophy, including definitions, goals, and educational guidance in Islam;
6. Formulating and planning the philosophy of Islamic science;
7. Investigating and studying, which led to the formulation of ways and content of disciplines and academic courses to be implemented at the university to integrate the various fields of science in all of the faculty and university;
8. Providing guidance for university to assess, investigate, and train scholars and intellectual leaders playing creative roles in restoring original Islamic *Tamaddun* at a reasonable place in the modern world;
9. Publishing the results of the investigation and assessment, which will be used all the time to be distributed to the Islamic world;

10. Establishing libraries of prime books in university which envisages intellectual and religious traditions of Islamic *Tamaddun* and the West as a way to achieve those goals and guidance;
11. Shaping and treating the Institute building, which will have more than 120,000 volumes of books, including manuscripts produced by previous Islamic scientists.

What is common in all such works is the zeal to show the existence of modern science in the Quran. This zeal is coupled with a desire to prove that the Quran is the Book of Allah, since it was impossible for anyone to know about this or that scientific fact in seventh century Arabia. They also use the well-known Quranic claim of its inimitability by restating it to mean that the Quran is inimitable because it contains precise scientific information that no human could have known in the seventh century and some of which remains unknown to humanity even now. By the end of the twentieth century, all verses of the Quran that could have been used to show its scientific content had received attention by scores of zealous writers. Lists of scientific verses of the Quran have been compiled; verses have been divided according to their relevance to various branches of modern science, such as physics, oceanography, geology, and cosmology.

Need to Change Curricula

In his Ph.D. thesis titled "*The Attitudes of Pakistani Science Teachers Toward Religion and Science*," a Pakistani scholar, Wali Mohammad Zaki of Harward University, mentions reasons for Muslim backwardness in the field of science. The results of Zaki's research are as follows:[11]

1. High school students in the United States understood the nature of the scientific enterprise and the methods and aims of science significantly better than did the high school teachers of West Pakistan.
2. The understanding of science, as well as attitudes towards science, were found to have a significant negative correlation with attitudes towards religion. Science teachers who had received their education in post-independence Pakistan were found to be more favourably inclined towards religion and less towards science, relative to those who had been trained in pre-partition days.
3. The denominational sects of Ahmadis and Protestants indicated a significantly more favourable attitude towards science than Sunnis.

4. Similar differences were found between regional cultures, with teachers belonging to Sind showing a significantly more favourable attitude towards science.
5. Science teachers with backgrounds in the biological sciences understand the nature of the scientific enterprise better than those in the physical sciences.

Although the thesis of Dr. Wali is the perception of a particular region of a Muslim country, it is the same thing that happens in almost all Muslim countries of the world. In research it has been concluded that the education systems and levels of education of Muslim countries are entirely mismatched with those of the western countries of Europe and the USA and the Asian countries of Japan and China. In a survey test of science and mathematics conducted by the National Institute of Psychology (NIP) in 1983 to compare the skills of schoolchildren from sixth to eleventh standard in various foreign countries against those of their Pakistani counterparts, the result was very alarming. The conclusions drawn by the NIP researchers were:[12]

1. The lowest scores of sixth class overseas children exceeded those of sixth, seventh, eighth, and ninth class Pakistani children in all cases. The highest score in mathematics of sixth class Japanese students (50.2) exceeded that of eleventh class Pakistani students (38.80). The report concludes that 'ultimately, many of our students of grade 11 remain less proficient in science and mathematics than grade 6 students in other countries.
2. Contrary to the belief that the standard of education is much higher in English-medium schools as compared to Urdu-medium schools, the NIP report found no major difference between the two in the learning of science and mathematics.

From the above conclusions, it may be observed that the standard of education in Muslim countries is much inferior in comparison to those of western countries. Adding further, that the medium of instruction is no bar for understanding science and technology, it is also noteworthy that the base knowledge of Muslim countries' students is very weak. And more interestingly, students in Muslim countries are more inclined towards religion than making them ready to face global scientific challenges.

The Pakistani institution, the Institute for Policy Studies (IPS), had set about the task of redefining science and drawing up guidelines for writing suitably Islamized science textbooks. The recommendations of IPS are as follows:[13]

(i) No phenomenon or fact should be mentioned without referring to the benevolence of Allah. For example, children should not ask 'What will happen if an animal does not take any food?' Instead, the following question should be asked: 'What will happen if Allah does not give the animals food?'

(ii) The effect must not be related to the physical cause. It is un-Islamic to teach that mixing hydrogen with oxygen automatically produces water. The Islamic way is this: when atoms of hydrogen approach atoms of oxygen, then by the will of Allah, water is produced.'

(iii) No law or principle should be named after scientists. For example, it is un-Islamic to speak of Newton's Laws, Boyle's Law, and so on, because this is tantamount to shirk (idolatry). Naming laws in this manner gives the impression that such laws were created rather than discovered by scientists.

(iv) The birth of all sciences should be traced to the Muslim period. Nuclear physics owes its origins to Ibn Sina, chemistry to Jabir Ibn Hayyan etc. The Greeks do not deserve credit because they knew nothing about experimental science.

Opposition of Theologians

Another most important reason for the backwardness of Muslim countries in terms of science and technology is undoubtedly the opposition of Muslim theologians in the modern world. Some believe that it is neither necessary nor desirable for Muslim science to 'catch up with the West', given the evil and godless nature of Western science. The old times were so much better; modernity brings nothing but corruption of the soul. Maulana Abul Ala Maudoodi, founder of the Jamaat-e-Islami, also bitterly criticises Western science. In a lecture on Islamic education, he stated that geography, physics, chemistry, biology, zoology, geology, and economics are taught without reference to Allah and his Messenger and are hence a source of straying from the truth.

Although the Holy Quran never prevents the use of the mind to study nature and explore inventions, theologians never go beyond the boundaries of religious learning. They thought that religion or religious knowledge would be polluted by mixing it with worldly scientific skill. They not only discouraged scientific knowledge in their institutions but bitterly opposed the teaching of science in their state. As the people were morally bound to the theologians, the rulers of those respective states also bowed before them. For example, an eminent biologist, Farukh Mohammad Ibrahim of Khartun University in Sudan, was jailed in 1990 for teaching Darwin's Theory of Evolution.[14]

Without a doubt, it was a political flop that had little to do with the Quran or Islam, and the Sudanese government was chastised for this harsh punishment. Negation is not the solution at all, but it requires deeper study and research to prove another's theory false. In this regard, they might have evolved a similar or parallel theory like Darwin's by research and observation in the light of their respective religion.

It shows that the scientific backwardness of Muslim states is proffered as proof that Islam is fundamentally retrogressive and incapable of sustaining a modern scientific culture. Moreover, it is a matter of great concern and rather shocking how far Muslim countries are drifting away from scientific study and research. In modern world, Islam is continuously refusing to consider purely secular knowledge under the pressure of theologians. Perhaps the theologians were afraid that secular knowledge would hamper their absolutism.

Debate over Islamic Scientific Traditions

Initially, the material translated from the Islamic scientific and philosophical traditions was deemed necessary for the development of science in mediaeval Europe. Later, a change took place in the European attitude towards Islamic scientific tradition. The colonisation of the Muslim world also contributed to European attitudes towards Islam and its tradition of learning, including the sciences. This colonizer-colonized relationship also affected European attitudes towards Islam and Muslims. In the meanwhile, the Muslim world itself was going through basic internal changes that drastically weakened the Islamic-scientific tradition and finally, choked it altogether.

Some scholars believe that Islam was never a forerunner of science and that all that Islamic civilization did was to "host" the Greek tradition it received through the translation movement for three centuries. During this period, science failed to take roots in the Muslim world because of fierce opposition from religious scholars. Greek science was then transferred to Europe, where it found its natural home and blossomed to become modern science.

On October 27, 1993, HRH Prince Charles delivered a lecture at Oxford, "If there is much misunderstanding in the West about the nature of Islam, there is also much ignorance about the debt our own culture and civilization owe to the Islamic world. It is a failure, which stems, I think, from the strait-jacket of history, which we have inherited. The mediaeval Islamic world, from Central Asia to the shores of the Atlantic, was a world where scholars and men of learning flourished. But, because we have tended to see Islam as the enemy of the West, as an alien culture, society, and system of belief, we have tended to ignore or erase its great relevance to our own history."[15]

On September 26, 2001, a businesswoman and historian, Carly Fiorina, chief executive officer at the time of Hewlett-Packard Corporation, gave a daring speech at a meeting of all the corporation's worldwide managers. In this meeting she said:[16]

> "There was once a civilization that was the greatest in the world. .... Its armies were made up of people of many nationalities, and its military protection allowed a degree of peace and prosperity that had never been known..... And this civilization was driven more than anything by invention. Its architects designed buildings that defied gravity. Its mathematicians created the Algebra and Algorithms that would enable the building of computers, and the creation of encryption. Its doctors examined the human body, and found new cures for disease. Its astronomers looked into the heavens, named the stars, and paved the way for space travel and exploration. Its writers created thousands of stories, the stories of courage, romance, and magic. Its poets wrote of love, when others before them were too steeped in fear to think of such things. When other nations were afraid of ideas, this civilization thrived on them, and kept them alive. When censors threatened to wipe out knowledge from past civilizations, this civilization kept the knowledge alive, and passed it on to others. While modern Western civilization shares many of these traits, the civilization I'm talking about was the Islamic world from the year 800 to 1600, which included the Ottoman Empire and the courts of Bagdad, Damascus, and Cairo, and enlightened rulers like Suleyman the Magnificent. Although we are often unaware of our indebtedness to this other civilization, its gifts are very much a part of our heritage. The technology industry would not exist without the contributions of Arab mathematicians...."

Organizational Efforts for Scientific Development

To deal with these complex global situations of the Muslim nations, certain consultative bodies have been formed. These include the Academy of Islamic Jurisprudence (AJC) established by the Organization of Islamic

Conference (OIC), the Islamic Organization for Medical Sciences, and the Research Council on Contemporary Issues established in Pakistan by certain Muslim scholars. Various committees and bodies were also formed by certain Muslim governments under their Ministries of Religious Affairs. These organizations have engaged a number of prominent Islamic scholars and scientists in deliberations about various aspects of new issues arising out of biomedical and other scientific research.[17]

Advancement and Devastation of the Muslim Nations

The Western countries developed themselves more rapidly, and there began a race to prove them more powerful. They manufactured weapons of mass destruction and sold them to developing countries, earning money from these sales. However, the Muslim nations hardly followed the West in their disastrous race. The Europeans and the United States generally imposed sanctions on the nations that had tried to prove them scientifically advanced.

Despite the many sanctions imposed by the West, the post-colonial era saw the emergence of numerous pragmatists as popular leaders of the Islamic world. Gamal Abdul Nasser, Ahmad Sukarno, Habib Bourguiba, Zulfiqar Ali Bhutto, Imam Khomeini, Saddam Hussain, Tayyeb Rajab Erdogan etc. were Muslim leaders who tried to make the Muslim nations scientifically strong. They were among those who called their peoples to action rather than admiration of Islam. The electoral defeats of the fundamentalist parties in several Muslim countries strongly suggest that most Muslims will not accept fundamentalist versions of the faith when there is an alternative.

Iran

Iran, traditionally known as Persia, had a very rich culture and civilization since the sixth century BCE. Its conquest by the Muslim Arabs in the seventh century CE left the most lasting influence. Its cities, like Tehran, Esfahan, and Shiraz, combine modern buildings with important landmarks from the past and serve as major centres of education, culture, and commerce. Iran fell under the sway of successive conquests of Turkish and Mongol conquerors until the rise of the Safavids, who introduced Shi'ism as the official creed in the early sixteenth century. With the fall of the Safavids in 1736, rule passed into the hands of several short-lived dynasties, leading to the rise of the Qajar line in 1796. Qajar rule was marked by the growing influence of the European powers in Iran's internal affairs.[18]

The Pahlavi line began to rule the country in 1925. The rule of the first Pahlavi monarch, Reza Shah, came about as a result of a military coup in 1921 supported by the British. Reza Shah crowned himself on April 25, 1926. Reza Shah was determined to modernise Iran, looking to imitate Ataturk's modernization of Turkey. Reza Shah's modernising efforts were mainly in the areas of the army, civil administration, finances, centralization of power, and the abolition of special rights to foreign powers. Reza Shah established the Tehran University in 1934, which was very modern and secular. Hamid Dabashi says of the university: "Tehran University would of course also function as the secular counterpart of religious seminaries in Qom and Mashhad—thus there were now two, diametrically opposed, educational systems, one almost exclusively under the control of the Pahlavi government, while the others thrived under a more or less autonomous clerical establishment."[9] In 1936, the veiling of women in public was banned by Reza Shah.

The second Pahlavi monarch, Mohammad Reza Shah, ascended the throne in 1941. Hamid Dabashi says of this transition: "The flirtation of Reza Shah with the Axis powers resulted in the Allied occupation of Iran soon after World War II began, and he was forced to abdicate in favour of his son, Mohammad Reza Shah, who was installed and supported in power by the Allied forces".[20] In 1943, Winston Churchill, Franklin Roosevelt, and Joseph Stalin met in Tehran, and they promised Iran national sovereignty. However, the Pahlavis always followed American guidelines to develop Iran in science and technology.

In 1963, the Shah government initiated the White Revolution, supported by President Kennedy, and implemented further secularization. It was challenged by the Iranian religious Shi'ite leader Sayyid Ruhollah Musavi Khomeini, also known as Ayatollah Khomeini (1902-1989), but he was arrested and exiled to Iraq. The United States has always supported Shah's development policies. In 1977, USA President Jimmy Carter proclaimed Iran "an island of stability in an otherwise troubled Middle East" on a visit to Iran.[21]

In 1978, the Shah government faced large public demonstrations by urban guerrilla movements and the clerics. These demonstrations led the army's to fire on protesters in the month of September, killing and wounding many. It is known as "Black Friday" in Iranian history. In October, Khomeini flew to France to direct the revolution from there. Actually, ill-planned efforts to modernise Iran led to widespread dissatisfaction and, after the movements against the Shah became violent, the Shah finally in January 1979 left the country for Egypt, where he eventually settled and died on July 27, 1980. This revolution brought a regime to power that uniquely combined elements, of a parliamentary democracy with an Islamic theocracy, run by the country's clergy.[22]

Under the new government of the Islamic Republic of Iran, the leader of the revolution, Khomeini, became the most powerful and Iran rose rapidly in terms of military power and prosperity. The only setback for Iran was the invasion of Saddam Hussein in September 1980, with the support of the US, Europe, and the Soviet Union. This Iran-Iraq War continued for eight-years which led to the drainage of the Iranian economy for a few years.

Iran has 523,000 active personnel under arms than any other country in the region, with 350,000 in the army, 18,000 in the navy, 30,000 serving in the air force and a further 125,000 in the Islamic Revolutionary Guards Corp (IRGC).[23] Years of international sanctions have meant it has been unable to source many weapons systems from abroad, forcing it to improvise and develop a substantial home-grown defence industry. Its armed forces are also seen as particularly strong in asymmetric warfare. It has also developed a very high-level educational system and conducted research in science and technology.

Iran has without doubt, the most advanced programme for the teaching of science and technology and the largest per capita number of scientists. It was the only country in the Muslim world where alternative technology was already beginning to be discussed. The government in Iran today wants very much to go back to implement the scientific and technological programmes.

Iraq

Iraq, a Muslim Arab nation, has a very glorious past. In modern times, Iraq developed more under the Ba'ath Party, and after its split in 1968, under the Arab Socialist Ba'ath Party from 1968-2003, as the Iraqi Republic (1968-92) and the Republic of Iraq (1992-2003). This period began with high economic growth and soaring prosperity and also as a military power. After the bloodless coup in 1968, Ahmed Hassan al-Bakr (1968-79) became the president, chairman of the RCC (Revolutionary Command Council-exercised both executive and legislative powers), secretary-general of the RL (Regional Leader of the Arab Socialist Ba'ath Party), and commander in chief of the armed forces. His assistant, Saddam Hussein, was appointed as deputy chairman of the RCC, deputy secretary-general of the RL, and vice president. However, in practice Al-Bakr was the *de jure* ruler and Saddam Hussein was the *de facto* ruler of Iraqi Republic.[24]

In response to the Iranian Revolution, several Iraqi Shias revolted against the Sunni-led government, which led to the collapse of the Ba'ath Party in certain areas of the country. The new Islamic Republic of Iran government also influenced many Shias to stand up against the Sunni-dominated government. In these critical situations, Al-Bakr resigned and Saddam Hussein (1979-2003) immediately succeeded him as president, chairman of the RCC, secretary-general of the RL, and commander in chief of the armed forces.[25] Saddam Hussein also served as Prime Minister of the Iraqi Republic from 1979-1991 and 1994-2003.

Iran-Iraq War (1980-1988): After the Iranian revolution, the spiritual leader of the Iranian government, Ayatollah Khomeini, proclaimed his policy of "exporting the revolution," and Iraq was high on the list of the countries whose governments were to be overthrown and replaced by a replica of the Islamic regime in Iran. In addition, Iran still occupied three small pieces of territory (Shatt al-Arab) along the Iran-Iraq border that were supposed to be returned to Iraq under the treaty of 1975. Saddam Hussein invaded Iran in September 1980 with the support of the US, Europe, and the Soviet Union.[26]

This Iran-Iraq War continued for eight-years that led to the setback of the Iraqi economy and worsened the standard of living of the Iraqi people. In this war, Iraq initially faced a setback, but during the latter phases of the war, the Iraqi Air Force established air superiority. In 1987, while Iraq raised an army of one million, Iran remained largely isolated from the international community. Iraq committed substantial resources in an attempt to develop or purchase other weapons of mass destruction (WMD), including biological and nuclear arms. The Iraqi government focused on Iraqi Kurdistan, which had revolted, and, as a result, Iraqi forces inflicted heavy casualties on Kurdish civilians by using chemical weapons. In April 1988, after a series of Iraqi military victories, a ceasefire was agreed between Iraq and Iran; the war was commonly considered *status quo ante bellum*.[27]

The war has been one of the most destructive conflicts of the late twentieth century. Hundreds of thousands had died, and large areas of western Iran and southeastern Iraq had been devastated. Even after the termination of the war, the dispute continued and each country supported opposition groups that worked against the rival's government. Iraq sheltered the Mojahedin-e-Khalq, an Iranian extremist group, and Iran lent support to various Iraqi Shi'i groups. Guerrilla attacks and terrorist incidents were frequent in the years after the war.

After the Iran-Iraq war, Saddam continued to purchase war weapons and also began its rearmament programme including expensive programmes to develop missiles and chemical, biological, and nuclear weapons. Meanwhile, Saddam began to make highly inflammatory speeches about the hostile Western attitude. In April 1990, he warned that if Israel ever again attacked Iraq (as that country had attacked and destroyed Iraq's Osiraq-Tammuz nuclear plant in 1981), he would retaliate with chemical weapons. This threat was later extended to include an attack by Israel on any Arab state, and Saddam soon began to suggest that Iraq's eventual goal was to defeat Israel and capture Jerusalem.[28] These declarations alarmed the Western powers more especially, the United States.

Iraq's Attack on Kuwait (1990) and Iraq War (1990-2003): Iraq was annoyed with the other Arab nations for their leaning towards Western powers. Tensions with Iraq grew further when several Gulf states, including Kuwait, exceeded their oil-production quotas set by the Organization for Petroleum Exporting Countries (OPEC). This resulted in a sharp drop in world oil prices, costing Iraq substantial amounts of income. Suspecting that the increase in oil production was prompted by Western pressure, Saddam Hussein criticized Kuwait and the United Arab Emirates.

The tension had another cause: Iraq had a historical claim to Kuwait's sovereignty dating to 1871, when the ruler of Kuwait was appointed sub-governor under Midhat Pasha. This claim had been pressed in 1961 during the Qasim regime but had met with strong resistance by other Arab states and the broader international community. It was also believed that an invasion of Kuwait would solve Iraq's two main problems—the urgent need for cash and the desire to control Bubiyan and Warbah.[29]

On 18 July, 1990, Saddam Hussein demanded that Kuwait repay Iraq for the oil it had (according to Saddam) stolen, and nullified Iraq's debt to Kuwait. The Kuwaiti leadership failed to respond, and on 2 August, 1990 the Iraqi military began the invasion of Kuwait. The invasion led to an international outcry; the United Nations, United States and the United Kingdom condemned the invasion and introduced sanctions against Iraq, and the Soviet Union and several Arab states also condemned the invasion. George H.W. Bush, President of the United States, demanded the immediate withdrawal of Iraqi troops from Kuwait and the restoration of the Kuwaiti government; Saddam responded by making Kuwait an Iraqi province.

A 28-member coalition, including several Middle Eastern countries and led by the United States, mobilised sufficient military and political support against Iraq. The coalition demanded that Iraq withdraw from Kuwait by no later than January 15, 1991, but the Iraqis seemed unconvinced. The coalition began air operations on January 17 and on

February 24, commenced a full-scale ground offensive on all fronts. The Iraqi military crumbled rapidly and capitulated after less than one week of fighting on the ground. The defeat compelled Iraq to withdraw from Kuwait and accept the Security Council resolutions.[30]

The Security Council lifted the embargo conditional on Iraq's accepting the demarcation of the Iraq-Kuwait border according to their bilateral agreement of October 1963, surrendering all its weapons of mass destruction, including missiles with ranges greater than 90 miles (150 km), and destroying the ability to create such weapons. The Security Council established a UN Special Commission (UNSCOM) to inspect and verify that Iraq was complying with the ban on Weapon of Mass Destruction.[31] Between 1991 and 1993, Iraq continually coordinated UNSCOM's search for WMD and argued that all of their proscribed weapons had been destroyed. However, UNSCOM deliberately insisted that the Iraqi regime was still concealing a small stockpile of prohibited items and technology.

Iraq accepted the Security Council's terms of the Oil for Food programme in 1996, and in December Iraq resumed oil exports. The first food and medicine shipments arrived in Iraq in March 1997. In February 1998, the Security Council again raised the ceiling on the permitted sales of Iraqi oil. Although, Iraq had no weapons of mass destruction and always coordinated with UNSCOM, the Western powers, including the United States and the United Kingdom, attacked Iraq in December 1998. United States used its dangerous Cruise Missile in this Operation Desert Fox on 1998. Although the United States used these missiles against Iraq in 1993 and 1996, but it was the most intense bombardment since the 1991 war.[32]

On 17th March, 2003, the United States and its allies declared an end to negotiations, and on 20th March, they launched the first in a series of precision air attacks on targets in Iraq. On 9th April, the resistance in Bagdad collapsed, and United States soldiers took control of the city. On that same day, British forces secured Basra, and Iraq's other major cities fell within days. Saddam Hussein went underground and most of his leaders and military men were killed in action. A new US-sponsored government was established in Iraq and Saddam Hussein was captured and executed by an Iraqi court on 30th December, 2006.

Afghanistan

Afghanistan, a Muslim country in the semi-fertile land and gateway of India, has always been strategically important for many countries, such as the Soviet Union, British India, and even the farthest countries like the United States. During the post colonial period in modern times, the Soviet

Union and the United States unsuccessfully tried to control Afghanistan. The Soviet Union signed a treaty with Afghanistan in 1978, and the two countries agreed to provide economic and military assistance. In Afghanistan in 1978, the communist government run by the head of the communist party, Nur Mohammad Taraki, was extremely unpopular with the Afghan people, and the Soviets attempted to bolster it with the treaty. Russia wanted to control over the Afghan-Russia border, a gateway to Asia, which was always considered important to its national security. Russia had long sought strong ties with China, having had interests there for centuries.

However, in September 1979, Taraki was overthrown and killed by members of the Afghan Communist Party who were unhappy with him. So, on December 24, 1979, Soviet Union invaded Afghanistan under the pretext of upholding the Soviet-Afghan Friendship Treaty. Russia begins a massive military operation, with Soviet leader Leonid Brezhnev sending in an estimated 280 transport aircraft and three divisions of almost 8,500 men each. Within a few days, the Soviets completed a special assault against Tajberg Palace and secured Kabul. The resistance was fierce but brief.[33]

After the Soviet Union's occupation of Afghanistan, the exiled leader of the Parcham faction of the Marxist People's Democratic Party of Afghanistan (PDPA), Babrak Karmal, was installed as Afghanistan's new head of government on December 27, 1979. However, the religious factions were annoyed with the new government as it was a puppet of the atheist Soviets. These religious leaders, along with other factions of rebel fighters from across the country, form the Mujahideen. They were backed by the West, who often referred to them as freedom fighters. They were funded by the United States and Pakistan. In 1980, US President Carter initiated a programme to arm the Mujahideen through Pakistan's intelligence service.[34]

Under President Reagan, US support increased. Named Operation Cyclone, it became the largest covert operation ever. Notwithstanding the new government in Kabul, the Mujahideen controlled most of rural Afghanistan. Spending his own money and using his connections, Osama bin Laden was also a major financier of the Mujahideen in the 80s. In 1987, the US introduced shoulder-launched anti-aircraft missiles, called Stingers, which allowed the Mujahideen to shoot down Soviet planes and helicopters. This turned the war, and the Mujahideen controlled most of the country by this point. As Gorbachev admitted defeat in 1988, Soviet Union soldiers began to withdraw, but 15,000 Soviet soldiers were killed in this war.[35]

In April 1988, elections were held and a coalition government was formed with incumbent leader Mohammad Najibullah as the head. Fifty seats are left open to encourage the Mujahideen to put down their arms. However, a new militant Sunni Islamist militant group, Al-Qaida, was formed in August 1988 amongst the Mujahideen. It was formed by Osama bin Laden, Abdullah Azzam, and other parties from the Soviet-Afghan war. The Mujahideen and Al-Qaida, who were still receiving funding from Washington and Islamabad, continued to fight against the Najibullah government, which is still being backed by Moscow. In 1990, the Afghan constitution was written and confirmed that the country was officially an Islamic state, and the last references to communism were removed. The head of state must now be from a Muslim Afghan family. Meanwhile, the Soviet Union officially collapsed and, without the support of Russia, Najibullah resigned in April 1992.[36]

With the support of Pakistan, some factions of the Mujahideen formed the Taliban in Afghanistan in August 1994. They recruited members from madrasahs across the border, but nearly all of the original leadership fought in the Soviet-Afghan War for the Mujahideen's Hezb-i-Islami Khalis or Harakat-i-Inqilab-i-Islami factions. Until November 1994-95, the Taliban conquered Kandahar City and controlled 12 Afghan provinces. In September 1996, they entered Kabul and established the Islamic Emirate of Afghanistan. They go on to establish an Islamic government based upon their interpretation of *Sharia* law.[37]

The World Trade Centre in the United States was attacked on September 11, 2001, causing the death of 2,977 victims. It was claimed that this terror attack was carried out by Al-Qaida chief Osama bin Laden. The United States and its allied forces attacked Al-Qaida training camps in Afghanistan in October-December 2001, but the Taliban government refused to hand over Osama bin Laden. The Taliban rulers flew from Kabul to the mountainous regions of Afghanistan. Western forces captured Kandahar, the last Taliban stronghold in the country's south. The Afghan Transitional Authority and the UN-backed International Security Assistance Force (ISAF) were set up after the Bonn Conference, in Germany, to maintain security in Afghanistan. However, endless fighting started between Taliban Guerillas and Western forces, causing much loss from both sides. At last, in November 2010, at a NATO Summit in Lisbon, a timetable for the handover of security control from the ISAF to Afghan forces by the end of 2014 was agreed. The transfer of security responsibilities from NATO to Afghan forces was completed on June 18, 2013. In between the main force Osama bin Laden was killed on May 2, 2011, by a US special operations unit after being tracked down living in a compound near Abbottabad, Pakistan. This led to the withdrawal of the US and UK forces from Afghanistan.[38]

On August 15, 2021, Taliban again formed their government and the United States evacuated diplomats from its embassy by helicopter. The NATO forces met a heavy loss and achieved nothing in the desert and hills, but they became successful in destroying the economy, warfare armaments, education, as well as science and technology. They pushed back the barren land of Afghanistan for many hundred of years.

Egypt

Like Iraq, Egypt also has a very glorious past. However, right from the ancient times after the emergence of Islam, Egypt was one of the strongest Muslim nations under the Caliphate and continued until modern times. During the post-colonial era, there was a coup in June 1953 which established the Arab Republic of Egypt (1953-1970). The coup leader was Mohammad Najib (1953-1956) an older officer who served as the figurehead for the Free Officers (previous government). After the coup, Mohammad Najib becomes president as Egypt is declared the Arab Republic of Egypt. Next year, Najib appointed his fellow coup leader, Gamal Abdel Nasser, as Prime Minister. During the rule of Najib, the Soviet Union continued to support Egypt and their relations continued to grow until 1955. The worst time of his regime was the attack by Israel on 28 February, 1955, in Gaza Strip, which was under the supervision of the Egyptian military.[39]

Gamal Abdel Nasser (1956-1970): In 1956, Gamal Abdel Nasser became the President of the Arab Republic of Egypt, ruling unchallenged until his death in 1970. During his presidency, Nasser was responsible for much of the progress in Egypt. Nasser gave women the right to vote and built the Aswan High Dam. Nasser introduced a socialist regime and moved Egypt closer to the Soviet Union. Under him, education and healthcare improved, but it was a repressive regime and the economy stagnated. Nasser was seen as one of the most important political figures in modern Arab history and politics. Under his leadership, Egypt nationalised the Suez Canal and a lot of other industries in 1956. Over the nationalization of the Suez Canal, Britain, France and Israel, invaded Egypt, but it failed due to the opposition of the United States. This diplomatic victory greatly enhanced President Nasser's standing at home and abroad. Nasser was even chosen as "Man of the Year" by "*Times*" magazine in 1956.[40]

Egypt's preferences and position began to take shape when Nasser started to develop good relations with the leaders of the Non-Aligned Movement countries like Jawaharlal Nehru and Josip Broz Tito. The period when the Movement appeared and Nasser's intention to benefit from joining this Movement can be considered an important development for Egypt's internal politics, as well as its access to Third World countries. From that point, Egypt's independent and impartial position and Nasser's importance in international relations would increase, especially in the late 60's. After 1959, Egypt began to pursue a more active policy in African politics by supporting their wars and, as a result, some twenty African countries declared independence in 1961.[41]

The biggest setback for Nasser was the Six-Day War with Israel, which broke out in June 1967. In this war, Israel wiped out almost all of Egypt's air and defeated Egypt, Jordan, and Syria, leaving it in control of Sinai up to the Suez Canal and Egyptian-controlled Gaza. After his defeat, Nasser wanted to resign, but people requested him to continue. He was a very popular leader, and during his presidency, Egyptian citizens managed to have housing, education, employment, healthcare, and nourishment, as well as other forms of social welfare. Egypt was capable of offering free education and healthcare to its own citizens but also to the citizens of other Arab and African countries. Nasser remained in office until his death from heart attack in Cairo on September 28, 1970.[42]

Anwar al-Sadat (1970-1981): After death of Nasser, his deputy, Anwar Sadat, took over the president's chair, and his position was approved by the people in the referendum held on October 15, 1970. Anwar Sadat approved of everything that Nasser did, and thus he was nicknamed "Yes President". However, despite this nickname, Sadat is seen as a person who exerted pressure to achieve what he wanted. Sadat made Egypt respectable for its military strength and political power in the Middle East. During his tenure, the Aswan High Dam is completed with Soviet assistance and has a significant impact on Egypt's irrigation, agriculture, and industry. President Sadat went to war with Israel in 1973 to reclaim the land lost in 1967. The Egyptian army was very successful at the start of the war, and their advance across the Suez Canal into the Sinai Peninsula surprised Israel and the rest of the world. This success made Sadat a hero in Egypt and, for a time, throughout the Arab world. He also reopened the Suez Canal in 1975 for the first time since 1967 war.[43]

In 1978, after the Camp David Accords, he visited Israel and signed a peace treaty in 1979 and got back to the Sinai Peninsula, but in return he also suspended itself from the Arab League. Over this peace treaty, many Islamists were very angry and planned to overthrow the Sadat government. The Egyptian government arrested many people thought to be part of the plot, but he missed a group of Islamists in the military who were led by Lieutenant Khalid Islambouli. This group assassinated Sadat on

October 6, 1981, during the victory parade celebrating the crossing into the Sinai Peninsula.[44]

Hosni Mubarak (1981-2011): After the death of Sadat, the Vice-President, Hosni Mubarak, became the President of Egypt in 1981. Mubarak's presidency lasted almost thirty years, making him Egypt's longest-serving ruler since Mohammad Ali Pasha. As President, Mubarak worked to improve Egypt's relations with other Arab countries. Mubarak believed that liberal methods should be applied in the economic and political life of Egypt. Mubarak himself gave more importance to the growth of his political power. In terms of foreign policy, the period of Mubarak continued the positive policy line in favour of Israel and the United States, launched during Anwar Sadat's period, trying to improve people's negative image regarding this issue. He also upheld the peace treaty that Egypt and Israel had agreed to in 1979.[45]

Politics in Egypt continued to follow authoritarian patterns, as Mubarak was re-elected to the presidency without opposition in 1987, 1993, and 1999, and although opposition candidates contested the 2005 election, he was re-elected that year as well. His National Democratic Party (NDP) continued to increase its majority of delegates in the People's Assembly in the elections held every five years. Under his rule, Egypt was one of at least six Arab countries developing domestic nuclear programmes to diversify energy sources, the International Atomic Energy Agency (IAEA) reports.[46]

In the 2010 parliamentary polls, the opposition party, the Muslim Brotherhood, failed to win a single seat, though it held a fifth of the places in the last parliament. This election effectively eliminated the opposition from the People's Assembly. It followed protests against alleged vote rigging. In January 2011, thousands of protesters crowded into the main streets of downtown Cairo, chanting slogans against corruption, political repression, and poverty. They protested unemployment, high prices, and the lack of democracy. As the demonstrations progressed, Egypt's established opposition groups increased their participation. Police Stations and NDP buildings throughout the country were attacked or set on fire by demonstrators. In Cairo, the NDP national headquarters was burned, and protesters fought running battles with security forces in the street. They demanded that Mubarak leave office.[47]

Mubarak stepped down after 18 days of demonstrations during the Egyptian Revolution of 2011. Mubarak resigned as president and transferred authority to the Supreme Council of the Armed Forces. He was accused of allegations of corruption and abusing power. On 2 June, 2012, an Egyptian court sentenced Mubarak to life imprisonment. But he was proven unguilty on March 2, 2017 by the Court of Cassation and released on March 24, 2017. Mubarak died on February 25, 2020, in Cairo.[48]

Mohammed Morsi (2012-13): Muslim Brotherhood candidate Mohammed Morsi narrowly won the presidential election in June 2012 and became the country's first democratically elected President. The Islamist-dominated constituent assembly approves a draft constitution that boosts the role of Islam and restricts freedom of speech and assembly. Throughout all of his presidential time, Morsi faced stiff opposition and protests. Because of the deterioration in the public services, calls for Morsi's resignation have increased. On July 3, 2013, the Egyptian military unseated Morsi. It suspended the constitution and created a new interim administration led by the head of the Supreme Constitutional Court, Adly Mansour. Morsi was placed under arrest and sentenced to death. On June 17, 2019, he collapsed while in court in Cairo and was pronounced dead shortly after. In May 2014, a presidential election was held, and former army Chief Abdul Fattah al-Sisi won the election.[49]

Libya

After World War II, Libya was ceded to France and the United Kingdom, and both countries linked it administratively to their colonies in Algeria and Tunisia. Libya became independent in 1951 and began to be ruled by the British-favoured Senussi dynasty under King Idris I. It was controlled by Saudi Arabia and endorsed by the United Nations. Under the monarchy of King Idris I, Libya was kept in total obscurantism while promoting British economic and military interests. Even though the oil reserves were discovered in 1959, the exploitation of wealth did not translate into benefits for the people. According to political analyst Thierry Meyssan, during the monarchy, the nation was mired in backwardness in education, health, housing, and social security, among others.[50]

Colonel Muammar al-Gaddafi (1969-2011): On September 1, 1969, Libyan Free Unionist Officers movement led by Colonel Muammar al-Gaddafi overthrew Senussi dynasty monarchy and deposed King Idris I. Colonel Muammar al-Qaddafi was nominated as the President of the Libyan Arab Republic on 13th September. They proclaimed true independence and removed the dominant foreign forces from the country. Libya was placed under the administration of the Libyan Revolutionary Command Council (RCC), a twelve-person body with Muammar Gaddafi at its head. Gaddafi was also named as Commander in Chief of the Armed Forces.[51]

Since Gaddafi took power, he undertook many changes in Libya. British and American troops withdrew from Wheelus Air Force Base and evacuated from Libya. Many foreign banks and hospitals in Libya were also nationalized. Italian and Jewish workers were expelled, and their properties were confiscated. The government also took over Libyan radio and television stations. As oil was the main resource in the hands of the leader of the newly proclaimed Libyan Arab Islamic Republic, Revolutionary Command Council (RCC) issued the first major law relating to the nationalisation of the oil industry, including British Petroleum Company and Bunker Hunt Oil. The new government used its oil income to boost redistributive measures among the population, generating a new model of economic and social development for the country. In 1973, Gaddafi nationalised all foreign-owned petroleum assets in the country.[52]

From 1974 onward, Gaddafi espoused a form of Islamic socialism as expressed in *The Green Book* ("The Solution to the Problem of Democracy: The Authority of the People", "The Solution to the Economic Problem: Socialism" and "The Social Basis of the Third Universal Theory"). This combined the nationalisation of many economic sectors with a brand of populist government ostensibly operating through people's congresses, labour unions, and other mass organizations. Although there was an unsuccessful military coup against Ghaddafi in 1975, and he severely punished revolutionaries and faced economic sanctions from the US, innovations along these lines continued to be introduced, including the *Jamahiriyyah* (a neologism implying a popular decentralised confederation) system of government in 1977.[53] These development policies resulted in a significant improvement in the living conditions of Libyans, who went from being one of Africa's poorest countries in 1969 to being the continent's leader in 2011.[54]

Gaddafi worked to strengthen ties with neighbouring countries such as Egypt, Morocco, Syria, Tunisia, and Chad, among others, as well as maintain close relations with countries like France and Russia. Gaddafi also connected with Latin American countries such as Venezuela and Cuba, which led him to cultivate an extensive network of contacts and an uncomfortable influence for Europe and the United States. Throughout Gaddafi's tenure, ambitious social programmes were launched in the areas of education, health, housing, public works, and subsidies for electricity and basic foodstuffs. Under Gaddafi, education and health care were free for all.

Gaddafi gave the world the *Great Man-Made River Project* to make water available to the whole country. It was an enormous irrigation plan throughout some of Libya's most arid areas. It is the largest irrigation project in the world to this day and covers 16,000 kilometres while supplying 70 per cent of Libya's fresh water. Moreover, Libya was a rich nation on its own. It did not have any debts.[55] These developmental policies led to a substantial improvement in the living conditions of Libyans, from being one of the poorest countries in Africa in 1969 to being the continent's leader in 2011.

In February 2011, after anti-government demonstrations forced Presidents Zine al-Abidine Ben Ali and Hosni Mubarak from power in the neighbouring countries of Tunisia and Egypt, anti-Gaddafi demonstrations broke out in the Libyan city of Benghazi. As the protests spread to many parts of the country, the Gaddafi regime attempted to violently suppress them, directing police and mercenary forces to fire live ammunition at protesters. As the opposition gained strength, international pressure on Gaddafi to step down increased. On February 26, the UN Security Council unanimously approved a measure that included sanctions against the Gaddafi regime, imposing a travel ban and an arms embargo and freezing the Gaddafi family's assets. On February 28, the United States announced that it had frozen $30 billion in Libyan assets linked to Gaddafi.[56]

As Gaddafi's forces advanced on Benghazi, the UN Security Council voted on March 17 to authorise military intervention to protect civilians (rebels) in Libya. The ensuing air campaign, led by the North Atlantic Treaty Organization (NATO), inflicted significant damage on pro-Gaddafi forces. They launched a bombing campaign between March and October 2011 to protect the civilians against attacks by the Libyan government forces. On April 30, a NATO air strike on Qaddafi's Bab al-Aziziyyah compound in Tripoli killed Gaddafi's youngest son, Sayf al-Arab, and three of Gaddafi's grandchildren. Gaddafi, reportedly in the targeted house at the time of the attack, escaped without injury. Following the air strike, NATO denied that it had adopted a strategy of trying to kill Gaddafi.[57]

On May 16, the International Criminal Court called for arrest warrants to be issued against Gaddafi, along with his son Sayf al-Islam and the Libyan intelligence chief, Abdullah Senussi, for ordering attacks on civilians during the uprising; the arrest warrants, for crimes against humanity, were issued on June 27. On October 20, 2011, Gaddafi's convoy was also bombed by NATO. He was captured, tortured and shot in the head by the rebels on October 20, 2011. This led to the end of Libya under Gaddafi.[58]

The West's assistance to Libyan rebels, as well as NATO's illegal intervention, was widely condemned around the world. Even the NATO, nations including the United States, have realised their mistake. Former US President Barack Obama admitted in 2016 that Libya was the biggest mistake of his presidency. Obama also made a searing critique of the former British Prime Minister, David Cameron and former French President Nicolas Sarkozy, for their roles in the bombing campaign in Libya. Moreover, Britain's Foreign Secretary Boris Johnson admitted that Libya is a result of the crime—a disaster, blunder, and 'big mistake' of the West.[59]

In Libya, there was never a mass movement on a national scale like in other countries, nor was there popular support to overthrow Gaddafi's government. The uprisings in Benghazi were the pretext for the United Nations Security Council and NATO to intervene to protect the civilians against the attacks by the Libyan government forces. Surely, Gaddafi was a better well-wisher of his people than the Western forces. It is believed that if NATO had not intervened in the internal affairs of Libya, the protesters could not have killed Gaddafi. Many agencies believe that the intervention of a multinational coalition led by NATO forces such as the United States, the United Kingdom and France on the pretext of protecting the civilians encouraged the rebels and eventually killed Gaddafi in October 2012. Actually, the West's objective was only to take Gaddafi out and take full control of Libya's natural resources—oil.

Turkey

For several centuries during the mediaeval period under the Osmanid Caliphate, Turkey kept itself at the forefront of history by the martial power and organisational superiority of its people. However, in modern times, they have been the object of great historical pressures because of their exposed and tempting weakness. After World War I, the Ottoman Empire shattered completely and gave forced birth to no less than seventeen independent nations in Europe, Asia, and Africa. Modern Turkey was one of them.

Through the efforts of Mustafa Kemal Ataturk, Turkey gained independence after a four-year long war of independence and a new shape of the Turkish Republic in 1923. The Kemalist Nationalists voted on the first constitutional document of the new Turkey on January 20, 1921. On October 29, 1923, the document establishing Turkey as a republic came into force. But the document known as the First Constitution of the Republic is dated April 20, 1924.[60] Mustafa Kemal Ataturk was the main leader of the revolution and the architect of the modern Turkish Republic. He founded a political party, Republican Peoples Party (CHP) in 1923, and ruled the country as President from 1923 till his death in 1938. Even after his death, his party continued to rule until 1950. In Turkey, nation-building practices emerged in the international state system as independent nation-states. The revolution and the independence of Turkey added a new chapter of development and progress to modern Turkey on secular and democratic lines.

The Turkish people, under Kemalist rule and ideology, were also free from the many foreign and domestic challenges to the establishment of a new nation. The secular intelligentsia, growing in prestige and numbers in urban centres, abandoned their acquiescent partnership with the civil service and began to search for means of more active participation. The new commercial and industrial middle class, flexing their muscle in an atmosphere of expanding domestic and international markets, became impatient with the paternalistic and authoritarian control of the state and demanded greater freedom of action. The population, which had become literate and urbanised and schooled in the ideology of a people's government, had desired proof of Ataturk's promise that the people's voice was the voice of God.[61]

The Kemalist Turkey, the first non-Western nation, was fully committed to a rational, secular, and democratic way of life. Turkey was the first non-Western nation to seek a new existence within the technical, political, and cultural mould of the West. Turkey has passed through all the phases of development. Turkey has tested all the easy and hard solutions in building a modern state. A systematic effort of development was started through mass education, conscription, and the like, to foster national integration and homogenization. Following the Kemalist ideology, Turkey spend the last three quarters of the twentieth century developing a new society patterned on the West.

The Republican People's Party abandoned its political monopoly in 1950, inaugurating a multi-party era. In the general election of 1950, the other parties like "Democrat Party", founded in 1946, "Liberal Republican Party", founded in 1930, and the "National Development Party" founded in 1945, contested the election. The Turks embraced the multi-party system with great, almost wild enthusiasm as the perfect answer to their quest for a true democracy. The Democrat Party won the election of 1950 with a thumping majority.[62]

The change of the Turkish Republic from a single-party dictatorship to a multi-party system elevated the common Turk from a nameless, voiceless being to an individual with a vote. The smooth transfer from a single-party dictatorship to a parliamentary democracy was hailed as a supreme achievement of the Turkish people. The world recognized, the new system in Turkey, as an instance of successful evolution from authoritarianism to representative government.

The Revolution took place in May I960, exactly ten years after the establishment of the multi-party system. The Turks had advanced one step closer to true democracy at this point, and it was a critical step because it brought the nation to the point of defining the relationship of man to man and man to society. This concern for individual rights is implemented in the new Constitution of the Republic, which was formulated after the coup of May 1960, and accepted by a popular referendum in May 1961. The Constitution of 1961, drafted by a group of Istanbul University professors, was submitted to a Constituent Assembly, convened by the National Unity Committee in Ankara on January 6, 1961. The Assembly completed its task on May 27, 1961. On July 9, 1961, the nation voted on it.[63] In this Constitution, more stress was given on the individual rights, personal freedom, liberties, and welfare of the individual and society. In this Constitution, the fundamental rights and duties were also widely elaborated.

Recep Tayyip Erdogan: In Turky, a new era of development has emerged under the leadership of Recep Tayyip Erdogan. He was a great sportsman, an Islamist, and a great political leader. In 1983, Erdogan and his old friends from the National Salvation Party established a party called Refah Party, which is also known as the "Explicit Islamic Party". In 1990, he was elected as a parliamentarian, and in 1994, he was elected as Mayor of Istanbul from the same party. In 1997, his Refah Party was forbidden as it was a threat to the secularism of Turkey. Erdogan gave a speech with a poem that stated, "The mosques are our barracks; the domes are our helmets; towers are our bayonets; and the believers are our soldiers." The poem is considered a violation of Kemalism. Thus, on December 12, 1997, Erdogan was imprisoned for four years because his poetry was considered an attempt to weaken the foundations of secularity in Turkey, but he was released after ten months in 1998.

After his release from jail, he and his party adopted a liberal policy and in 2001 formed another political party, "Adalet ve Kalkynma Partisi" (AKP; Justice and Development Party). In those times, there was a ban in Turkey on religiously based political parties, so the AKP had its main agenda of liberal democracy and secularism, and this party had started a movement for the same. Although its origins can be traced back to the Turkish Islamist movement, and its leader and founder, Recep Tayyip Erdogan, was previously known for his Islamic leanings.[64] Recep Tayyip Erdogan's party, AKP, won the parliamentary elections in 2002, but Erdogan was legally barred from serving in parliament or as prime minister because of his 1998 conviction. A constitutional amendment in December 2002, however, effectively removed Erdogan's disqualification. On

March 9, 2003, he won a by-election and a few days later he was asked by President Ahmet Necdet Sezer to form a new government. Erdogan took office on May 14, 2003.[65]

After the 2001 financial crisis, the Turkish economy grew rapidly, and the country's level of economic development is higher than that of many countries which have consolidated their democratic regimes in the past two decades. During five-year rule of the AKP, the Islamist ideology fully developed in the country and challenged the Kemalist secular ideology. He is an official government that succeeded in transforming Turkish society, which was once attached to secularism, into a society that is now attached to Islamism.

Despite the strong opposition of the Kemalist group, elections were held in Turkey in 2007, and the AK Party won a landslide victory, garnering 46.5 per cent of the popular vote and 341 seats. It was the first time in fifty years that a ruling party was returned to power with a bigger share of the vote.[66] The party under the leadership of Recep Tayyip Erdogan won a strategic position, namely the position of President, which occupied by Abdullah Gul, while the Prime Minister was occupied by Recep Tayyip Erdogan. The rapid progress of the Turkish state under Erdogan's leadership as the Prime Minister made Turkey now respected and became one of the strongest states in Europe. It also made his own name soar and he became one of the best leaders in the world.

The glory of the AKP was again repeated in the 2011 general election, and the AKP won the trust of the citizens to re-win, and Erdogan served as Prime Minister for the next period. When his term as Prime Minister of Turkey was completed in 2014, Recep Tayyip Erdogan then tried to run for President of Turkey with the support of AKP. On August 10, 2014, Erdogan was successfully elected as the President of Turkey through general elections. In accordance with the constitutional amendments of 2007, the 2014 election was the first time that the President was elected directly, rather than by the parliament. It was certainly an extraordinary achievement for Erdogan. The success of Erdogan won the hearts of the people, not only in economic and secular programmes but mainly because of his Islamization programme. But in the general parliamentary election of June 2015, the AKP indeed lost its parliamentary majority after 13 years of single-party rule. However, after a few months later, AKP was able to prove its parliamentary majority.[67]

Erdogan survived a violent coup attempt in the summer of 2016. On the night of July 15, a small number of military personnel occupied streets in Ankara and Istanbul and seized facilities, including television stations and bridges. Erdogan and the AKP used social media to rally support. The coup plotters were soon overpowered by loyal military units and civilians, and the government quickly regained control. Nearly 300 people, mostly civilians, were killed in confrontations during the coup. In April 2017, some changes were made to the Constitution that abolished the post of prime minister and empowered the president as the executive head of government. In July 2018, Erdogan was re-inaugurated as the President of Turkey. The votes that he achieved were from the majority of conservative Muslims. Aside from that, Turks had entered a new era in the twentieth century.[68]

As a Muslim and a public figure, Erdogan is known as a religious leader. His communication, closeness, and leadership attitude made people love Erdogan as their leader. He began his political career by arousing Islamic values as a guide to the life of Turkish society. Erdogan provided scholarships to students at the beginning of the school year. He eliminates the cost of education where the costs are borne by the government, returning the old habit of teaching the Quran and *Hadith* in schools that have been long eliminated, requiring Islamic religious education from primary and secondary levels to twelve grade levels, and teaching Arabic as a second language as well. These were done by Erdogan to enable students to understand the Quran.

Erdogan promises a powerful Turkey, projecting it as one of the greatest powers in the Middle East, capable of competing with Europe and other world powers. Erdogan's foreign policy was to balance the Middle East and Western countries. In defence and security cooperation, Turkey and the United States focus on the military sector. Some of the US policies in terms of Turkish defence strategies include contributing to the defence of safeguards in guarding Turkey's border areas and supporting strong army reserves to be assigned to areas under threat in domestic or regional regions of Turkey. With this assistance, the United States also supports Turkey in its efforts to join the European Union. Adequate Turkish military strength and a strong defence base serve as a basis for participating in joining the European Union.

In the economic field, Erdogan cooperates with various parties, especially Arab countries. Erdogan raised the level of new prosperity in the economic sector and respected international policy. Erdogan continued to hit the Turkish economy so hard that it experienced a resurgence. Since taking over as Prime Minister in 2002, Erdogan immediately started economic reforms to expel the country from the crisis. He also collaborated with Iran, as a neighbouring country, which has the second largest natural gas reserves in the world. Because, Turkey does not have abundant natural gas reserves, while natural gas is the main energy source that contributes most to the rate of the Turkish economy. Turkey's Gross Domestic Product then reached trillions of dollars, so it became the sixteenth country in the world's economic power. In 2020, Turkey plans to be ranked tenth in the world. Economic growth averaged 9 per cent per year, making Turkey the seventh country in economic power in Europe.

Erdogan has revealed himself as an increasingly authoritarian leader, convinced that his electoral victories legitimise the indiscriminate use of political power. At the same time, the AKP has become a party extremely focused on Erdogan, in which loyalty to the leader has become a promotion criterion. By entering into force only in 2019 and changing the limit of presidential terms, the new Constitution will allow Erdogan to remain in office until 2029. This is a type of government longevity that is unprecedented in Turkey, which began with a proposal for a moderate and liberal political orientation and gradually became conservative and authoritarian.

In Turkey, an incredible amount of intellectual activity geared towards reviving the past glory and scientific inquiry and research has been going on in the last few decades. The most important journal which is being published in Turkey on this issue, called "Science and Technology", is being published by very devout Muslims, who are extremely interested in the Islamicity of Islamic science, and certainly the Turks will be able to make some major intellectual contributions in the future to this field.

Syria

Syria is an Islamic country located on the east coast of the Mediterranean Sea in south-western Asia, with its capital being Damascus. Its area includes territory in the Golan Heights that has been occupied by Israel since 1967. In ancient times, it had a very golden past, a vast land having a strip of fertile land lying between the eastern Mediterranean coast and the desert of northern Arabia. In 1946, Syria concluded a treaty with France, ending French rule in Syria, and the French troops were withdrawn. After Syria gained its independence in 1946, political life in the country was highly unstable, owing in large measure to intense friction between the country's social, religious, and political groups. There was no stable and powerful government in Syria until 1970.

In 1970, Syria came under the authoritarian rule of President Hafez al-Assad through a coup. The Assad government's most important goals included achieving national security and domestic stability. However, his foremost goal was to recover the Syrian territory, Golan Heights, which was lost to Israel in 1967, in the Six-Day War, in which Egypt, Jordan, and

Syria were defeated by Israel. Assad committed his country to an enormous arms buildup, which put severe strains on the national budget. To fulfil the foremost goal, Syria and Egypt launch attacks against Israeli forces in the Golan Heights and the Sinai, respectively, but Syria fails to retake the Golan Heights. Hostilities end with a cease-fire agreement.[69]

After Assad's death in 2000, his son Bashar al-Assad became President. Despite some early steps towards political reform, Bashar al-Assad ultimately continued his father's authoritarian style of govérnment, using Syria's powerful military and security services to quash political dissent. Long-suppressed internal tensions led to the outbreak of the Syrian Civil War in 2011. In the Civil War, pro-democracy demonstrations erupted in the southern city of Deraa, inspired by uprisings in neighbouring countries against repressive rulers. Protesters demanded an end to the authoritarian practices of the Assad regime. The Syrian government used violence to suppress demonstrations, making extensive use of police, military, and paramilitary forces. Opposition militias formed in 2011, and by 2012, the conflict had erupted into a full-fledged civil war across the country.

President Assad vowed to crush what he called foreign-backed terrorism. Foreign powers began to take sides, sending money, weaponry and fighters, and as the chaos worsened, extremist jihadist organisations with their own aims, such as the Islamic State (IS) group and Al-Qaeda, became involved. It followed a lot of bloodshed and devastation. The government's key supporters have been Russia and Iran, while Turkey, Western powers, and several Gulf Arab states have backed the opposition to varying degrees during the conflict. The Russian Fereration, which had military bases in Syria before the war, launched an air campaign in support of President Assad in 2015 that has been crucial in turning the tide of the war in the government's favour. The Russian military says its strikes only target terrorists, but activists say they regularly kill mainstream rebels and civilians. Iran also deployed hundreds of troops and spent billions of dollars to help President Assad. Iraq, Afghanistan, and Yemen have also fought alongside the Syrian army.[70]

The US, UK, and France initially provided support for what they considered moderate rebel groups. But they have prioritised non-lethal assistance since jihadists became the dominant force in the armed opposition. A US-led global coalition has also carried out air strikes and deployed special forces in Syria since 2014 to help an alliance of Kurdish and Arab militias called the Syrian Democratic Forces (SDF) capture territory once held by IS militants in the north-east and stop the jihadist group from rebuilding. Turkey was a major supporter of the opposition, but its focus was on using rebel factions to contain the Kurdish YPG militia that dominates the SDF, accusing it of being an extension of a banned Kurdish rebel group in Turkey. Saudi Arabia, which was keen to counter Iranian influence, armed and financed the rebels at the start of the war, as did the kingdom's Gulf rival, Qatar.[71]

After dozens of Turkish soldiers were killed, the Russian-backed Syrian government forces came into direct confrontation with Turkey. However, the confrontation ended after a cease-fire was brokered by Turkey and Syria's ally, Russia. The government has regained control of Syria's biggest cities, but large parts of the country are still held by rebels, jihadists and the Kurdish-led SDF. There have been no shifts in the front lines for two years. The last remaining opposition stronghold is in the north-western province of Idlib and adjoining parts of northern Hama and western Aleppo provinces. The region is dominated by a jihadist alliance called Hayat Tahrir al-Sham (HTS), but is also home to mainstream rebel factions.

Before the war, of course, Syria had a completely different identity. Tourists flocked to the country for its barbecue, its culture, and its rich history, eager to explore several of the oldest cities in the world. In today's issue, we are reflecting on Syria before the war. But the eleven years of war have inflicted immense suffering on the Syrian people. The Syrian Observatory for Human Rights (SOHR), a UK-based monitoring group with a network of sources on the ground, had documented the deaths of 4,94,438 people by June 2021.[72]

It said at least 159,774 civilians had been killed, with the Syrian government and its allies responsible for most of those deaths. More than half of Syria's pre-war population of 22 million had fled their homes. Some 6.9 million people were internally displaced, with more than two million living in tented camps with limited access to basic services. Another 6.8 million were refugees or asylum-seekers abroad. Lebanon, Jordan, and Turkey, which housed 84 per cent of the refugees, have struggled to cope with one of the largest refugee exoduses in recent history.[73]

According to the facts stated above, it was Europeans and the United States who provoked the rebellious forces and supplied them with heavy weapons against Muslim developing Muslim countries in order to crush the emerging Muslim powers. They also attacked Muslim countries on the pretext of saving the innocents and completely ruined the country's political authorities. It is also very strange that they did not even spare the Islamic rebellious groups whom they had earlier supported with arms and money and brutally butchered on the pretext of terrorism.

Pakistan

Pakistan came into existence after the partition of India in 1947, but India and Pakistan have been in conflict over several issues, including the dispute over the territory of Jammu and Kashmir. However, Pakistan lags far behind its counterpart India in terms of economic resources and infrastructure, science and technology, and arms and ammunition. Undoubtedly, the United States helped Pakistan with armaments in wars against India and China. As for the economy, Zulfiqar Ali Bhutto tried his level best to improve the financial position of Pakistan. However, Pakistan was always second to India due to its limited natural resources and some of the unfertile lands. To keep up with Indian development, Pakistan has always relied on the United States for aid and advancement.

Pakistan developed its own missiles and nuclear weapons in the late twentieth century with the efforts of its world famous Physicist, Abdus Salam, (1926-1996). Professor Abdus Salam was the first Muslim and the first Pakistani nuclear physicist, who was the co-recipient with Steven Weinberg and Sheldon Lee Glashgow of the 1979 Nobel Prize for Physics for their work in formulating the electroweak theory, which explains the unity of the weak nuclear force and electromagnetism. Abdus Salam, the only Muslim to have won the Nobel Prize in physics, was asked "What happened to Islamic Science?" He said, "Nothing. Instead, what we cultivated in Isfahan and Cordoba is now being cultivated at MIT, Caltech and at Imperial College, London. It is just a geographical translation of a place."[74]

Throughout his life, Salam was a dedicated Muslim. He listened to the Koran on repeat while he worked in his office in his London home. He never saw his religion as a barrier to his science. In fact, he saw them working together and claimed to colleagues that many of his ideas came to him from God. He was striving for a unified theory that would explain all of particle physics, which was in line with his religious beliefs. In 1961, he established Pakistan's space programme, while during the early 1970s, Salam was, controversially, involved in Pakistan's efforts to build a nuclear weapon. But after Prime Minister Zulfikar Ali Bhutto passed the law against the Ahmadiyya Muslims in 1974, Salam's involvement with the country's administration finally diminished.[75]

Salam is intimately familiar with the Golden Age of science in Islam and contends that modern science is not a creation only of the Western, Judeo-Christian tradition, as is often claimed, but that contemporary scientific advances have their roots in innovations and discoveries made earlier in Islamic lands. In making the claim, he is neither argumentative nor contentious but rational and scholarly, quoting extensively from eminent historians and researchers. Thus, we have Briffault testifying, "The Greeks systematized, generalized, and theorized, but the patient ways of detailed and prolonged observation, and experimental inquiry were altogether alien to the Greek temperament .... What we call science arose as a result of new methods of experiment, observation and measurement, which were introduced into Europe by the Arabs. (Modern) science is the most momentous contribution of Islamic civilization." These thoughts are re-echoed by George Sarton, the great historian of science: "The main, as well as the least obvious, achievement of the Middle Ages was the creation of the experimental spirit, and this was primarily due to the Muslims down to the twelfth century."[76] With his International Centre for Theoretical Physics in Trieste, Abdus Salam has relit the torch that burned so brightly with such a clear light in Bagdad, Cairo, and Cordoba one thousand years ago.

Pakistan began development of nuclear weapons in January 1972 under Prime Minister Zulfikar Ali Bhutto, who delegated the programme to Munir Ahmad Khan, who made the device ready, and its first test was held on May 28, 1998, a few weeks after India's second nuclear test. Pakistan's nuclear weapons development was in response to the loss of East Pakistan in 1971's Bangladesh Liberation War. In this project, many other scientists, such as Dr. Abdul Qadeer Khan, Dr. Ishrat Husain Usmani, and Dr. Abdus Salam Khan, played an important part. Pakistan is now one of nine states to possess nuclear weapons. In May 1999, during the anniversary of Pakistan's first nuclear weapons test, former Prime Minister of Pakistan Nawaz Sharif claimed that Pakistan's nuclear security is the strongest in the world. According to Dr. Abdul Qadeer Khan, Pakistan's nuclear safety and nuclear security programme are the strongest in the world.

It is a truly unique case of Pakistan in which Western powers never subdued the development process, not even its nuclear programme, but instead assisted in uranium, plutonium, and technology arrangements. In the West, politics became more powerful than religion. Actually, the West never thought of Pakistan as a danger to them, unlike other Muslim countries in the world. They deliberately wanted to create a strong power against India, which was the fastest emerging power in terms of nuclear energy and warfare armaments. They created Pakistan as a barrier between India and the West.

Palestine

The first Caliph, Abu Bakr (632-34), laid the successful unification of the Arabian Peninsula. However, before completion of his deram, he died and the conquests he started were carried on by his successor, the Caliph Umar I (634-644). The Islamic forces sent by him won a bloody victory on July 30, 634, at a place in southern Palestine. All of Palestine then lay open to the invaders. In the meantime, the emperor Heraclius dispatched a large army against the Muslims in 636. The decisive battle that delivered Palestine to the Muslims took place on August 20, 636. Jerusalem and Caesarea were also surrendered to the Muslims in 638 and 640. Now, Palestine, and indeed all of Syria, was then in Muslim hands.[77]

Under Muslim rule, Palestine developed economically, culturally, and educationally. It was famed for its trades, culture, and scholarship. On visiting the Temple Mount area, which Muslims came to know as Al-$aram al-Sharif and finding the place suffering from neglect, Caliph Umar and his followers cleaned it with their own hands and declared it a sacred place of prayer, erecting there the first structure called Al-Aqsa Mosque.

Palestine suffered a disastrous fate at the hands of both Mongols and Crusaders at the end of the eleventh century and continued to the twelfth century. Jerusalem's flourishing character prior to the crusades is noted by the traveller Nasr Eddin Khusraw in 1047. There, he notes, things were cheap and plentiful; beautiful markets and high buildings everywhere; a great number of craftsmen; and each craft had its own market. About a hundred thousand people lived in the city, which also had a great hospital with rich waqfs (endowments) dedicated to it, from which medicines for its numerous patients were dispensed and salaries for doctors were paid, and where medicine was taught.[78]

In the same eleventh century, under the Seljuks, just on the eve of the crusades, Jerusalem sheltered a great variety of cultural activities. Great scholars from both the east and west of the Muslim lands made the city their destination, many settling there; both the city's scholars and the visitors participated in a rich cultural life. Ibn al-Arabi gives a vivid picture of such activities in the city, a meeting place of scholars from Khurasan in the east to Al-Andalus (Muslim Spain) in the west. Ibn al-Arabi was impressed by the *majlis* (study circles) and scholarly disputations.[79]

When the crusaders arrived in July 1099, they inflicted on Jerusalem its worst woes in history. Storming through the city walls, they slaughtered the whole Muslim population. A contemporary, Abbot Raymond of Agiles, present during the dramatic moments, wrote with glee:[80]

> "When our men took the main defences, we saw then some astonishing things amongst the Saracens. Some were beheaded, and that's the least that could happen to them. Others were pierced through and so threw themselves from

> the heights of the walls; others after having suffered at length were thrown into the flames. We could see in the roads and in the places of Jerusalem bits and pieces of heads, hands, and feet. Everywhere we could only walk through cadavers. But all that was only little..." The Abbot's description moves onto the Mosque of Umar, where: "There was so much blood in the old temple of Solomon that dead corpses swam in it. We could see hands floating and arms that went to glue themselves to bodies that were not theirs; we could not distinguish which arm belonged to which body. The men who were doing the killing could hardly bear the smoke from the corpses."

Through the efforts of Nooruddin Zengi and Salahuddin Ayyubi, the tide was turned. On July 4, 1187, he dealt the Crusaders a crushing blow at the decisive Battle of Hittin. Most of Palestine came once again under Muslim rule. Further attempts by the Crusaders, to regain control of Palestine, proved ineffective, primarily because of incessant quarrels among the Crusaders themselves.[81]

Around 1900, the Jews' migration to Palestine and purchase of lands rose vehemently, and the Arabs feeling the unknown danger, opposed their move. The population of Palestine, predominantly agricultural, was about 690,000 in 1914 (535,000 Muslims; 70,000 Christians, most of whom were Arabs; and 85,000 Jews). The Jewish population was increasingly composed of people born in Israel itself, although millions of immigrants have arrived since the founding of the State of Israel in 1948. The Arab majority resident in the West Bank and the Gaza Strip have strongly opposed Israeli control and feared an eventual annexation of the West Bank and Gaza by Israel.

After World War I and the end of rule by the Ottoman Empire, Palestine was adopted or mandated by Great Britain, roughly comprising present-day Israel and the West Bank. The new administration proceeded to implement the Balfour Declaration (establishment in Palestine of a national home for the Jewish people on the understanding that "nothing shall be done which may prejudice the civil and religious rights of existing non-Jewish communities in Palestine), announcing in August 1920 a quota of 16,500 Jewish immigration for the first year. It was the first move to establish a Jewish home in Palestine.[82]

The arrival of more than 18,000 Jewish immigrants between 1919 and 1921 and land purchases in 1921 by the Jewish National Fund (established in 1901), which led to the eviction of Arab peasants (fellahin), further aroused Arab opposition that was expressed throughout the region through the Christian-Muslim associations. Arab organisations in the 1920s opposed Jewish immigration, including the Palestine Arab Congress, Muslim-Christian associations, and the Arab Executive. The Nazi's accession to power in Germany in 1933 and the widespread persecution of Jews throughout central and eastern Europe gave a great impetus to Jewish immigration, which jumped to 30,000 in 1933, 42,000 in 1934, and 61,000 in 1935. By 1936, the Jewish population of Palestine had reached almost 400,000, or one-third of the total.[83]

In August 1945, United States President Harry S. Truman requested that British Prime Minister Clement Attlee facilitate the immediate admission of 100,000 Jewish Holocaust survivors into Palestine, and in December, the US Senate and House of Representatives asked for unrestricted Jewish immigration to the limit of the economic absorptive capacity of Palestine. Truman's request signalled the US entry into the arena of powers determining the future of Palestine. The question of Palestine, now linked with the fate of Holocaust survivors, has once again become the focus of international attention.[84]

The major issue between 1945 and 1948 was, as it had been throughout the mandate, Jewish immigration to Palestine. British Foreign Secretary Ernest Bevin opposed Jewish immigration and the foundation of an independent Jewish state in Palestine, but Truman was determined to ensure that Jews displaced by the war were permitted to enter Palestine. On August 31, 1947, a majority report of the UN Special Committee on Palestine (UNSCOP) recommended that the region be partitioned into an Arab and a Jewish state, which, however, should retain an economic union. Jerusalem and its environs were to be international. This recommendation was substantially adopted by a two-thirds majority of the UN General Assembly in Resolution 181, dated November 29, 1947, against the wishes of the majority of its inhabitants (in 1946 there were 12,69,000 Arabs and 6,78,000 Jews in Palestine). The issue was resolved in 1948 when the British mandate collapsed under the pressure of force and diplomacy. The Zionists welcomed the partition proposal both because it recognized a Jewish state and because it allotted slightly more than half of (west-of-Jordan) Palestine to it.[85]

On May 14, 1948, the last British High Commissioner, General Sir Alan Cunningham, left Palestine. On the same day, the State of Israel was declared and, within a few hours, won *de facto* recognition from the United States and *de jure* recognition from the Soviet Union. The violent birth of Israel led to a major displacement of the Arab population, who either were driven out by Zionist military forces before May 15, 1948, or by the Israeli army after that date, or fled for fear of violence by these forces. More than 400 Arab villages disappeared, and Arab life in the coastal cities (especially Jaffa and Haifa) virtually disintegrated.[86]

The centre of Palestinian life shifted to the Arab towns of the hilly eastern portion of the region, which was immediately west of the Jordan River and came to be called the West Bank. Nearly 14,00,000 Arabs lived in Palestine when the war broke out, but more than 6,00,000 or 7,00,000 Arabs were displaced from their homes. Some 2,76,000 moved to the West Bank; between 1,60,000 and 1,90,000 fled to the Gaza Strip (40 km long and 6-8 km wide, it became one of the most densely populated areas in the world). More than one-fifth of Palestinian Arabs left Palestine altogether. About 1,00,000 of these went to Lebanon, 1,00,000 to Jordan, between 75,000 and 90,000 to Syria, 7,000 to 10,000 to Egypt, and 4,000 to Iraq.[87]

The Palestenians became refugees in their own country, and landlords became labourers in the factories of Israel. It was an international injustice by the most civilised countries in the world. Although the Palestinians tried to save their home hearths and sacrificed their lives and property in various civil wars, revolts, and riots, due to international conspiracy and the leading role of the United States, they lost their efforts and the majority was defeated by the minority. Israel, with the support of the United States, strategically and militarily became more powerful and slowly and gradually occupied more Palestinian lands, flouting the land partition of 1948 by the United Nations. Even the United Nations openly supported almost all of Israel's moves. Palestinians were left homeless and at the mercy of the rest of the Arab countries.

NOTES

1. Iqbal, Muzaffar, *op. cit.*, p. 192.
2. *Ibid.*, p. 164.
3. *Ibid.*, p. 180.
4. *Ibid.*, pp. 195-96. The creation of the heavens and earth are most often cited in the verses (Quran 21:30 and Quran 41:11). The former states, "Do the disbelievers not see that the heavens and the earth were joined together, then We clove them as under and We created every living thing out of water. Will they then not believe"? The latter reads, "God turned toward the heaven and it was smoke (The smoke refers as predominantly gaseous state of

the material that composes—the universe, which—obviously corresponds to the concept of the primary nebula put forward by modern science)...." Other verses pertaining to creation mention "six days" during which the heavens and the earth and all that is between them were created by God. The six days are shown to mean six indefinite periods of time. [Maurice Bucaille (2000). *The Bible, the Quran, and Science: The Holy Scriptures Examined in the Light of Modern Knowledge*, New Delhi. pp. 149-54.]

5. Keith, L. Moore (1982). *The Developing Human: Clinically Oriented Embryology*, Philadelphia. p. 10.
6. Iqbal, Muzaffar, *op. cit.*, p. 195.
7. *Ibid.*, pp. 193-94.
8. *Ibid.*, p. 194.
9. *Ibid.*, p. 209.
10. *https://www.academia.edu/37479560/CHAPTER_III_BIOGRAPHY_OF_SYED_MUHAMMAD_NAQUIB_AL_ATTAS_AND_HIS_THOUGHT_TOWARD_ISLAMIZATION_OF_KNOWLEDGE.*
11. Hoodbhoy, *op. cit.*, pp. 39-40.
12. *Ibid.*, p. 40.
13. *Ibid.*, pp. 54-55.
14. *Ibid.*, p. 47.
15. Al-Hassani, *op. cit.*, p. 11.
16. *Ibid.*
17. Iqbal, Muzaffar, *op. cit.*, p. 221.
18. *https://www.britannica.com/place/Iran.*
19. Hamid Dabashi (2007). *Iran: A People Interrupted*, New York, p. 121.
20. *Ibid.*, p. 123.
21. *Ibid.*, p. 157.
22. *Ibid.*
23. *https://www.forbes.com/sites/dominicdudley/2018/02/26/ten-strongest-military-forces-middle-east/?sh=2d7c597f16a2.*
24. Coughlin, Con (2005). *Saddam: His Rise and Fall*, New York, pp. 53-58.
25. Tripp, Charles (2010). *A History of Iraq*, Cambridge, pp. 212-13.
26. *Ibid.*, pp. 224-25; *see also https://www.britannica.com/place/Iraq/Iraq-under-Saddam-Hussein.*
27. *Ibid*, pp. 233-38.
28. *https://www.britannica.com/place/Iraq/Iraq-under-Saddam-Hussein.*
29. Roberts, Paul (2005). *The End of Oil: On the Edge of a Perilous New World.* Boston, p. 105.
30. Fisk, Robert (2005). *The Great War for Civilisation: The Conquest of the Middle East.* London. pp. 646-647. (After Iraq's defeat, the heavy hand of Saddam fell on Kurds and other Shi'is as they were involved in anti-state activities during the war. Even after Iraq's defeat, Shi'is arose against the regime, but their rebellion was put down with great brutality. The regime allegedly killed scores of prominent Shi'i religious and political leaders

and arrested and imprisoned thousands of others whom they accused of sedition. Shi'is and Kurds suffered the regime's greatest wrath. Even the seditious Sunnis, including members of Saddam's own family, dealt with brutality.)

31. *https://www.britannica.com/place/Iraq/Iraq-under-Saddam-Hussein.*
32. *Ibid.*
33. *https://inews.co.uk/news/world/russia-afghanistan-why-invade-soviet-union-invasion-1979-timeline-what-happened-1156206.*
34. *Ibid.*
35. *Ibid.*
36. *Ibid.*
37. *Ibid.*
38. *Ibid.*
39. ILIRIA International Review, Vol. 6, No. 1, 201, (Islam Islami, *Political History of Modern Egypt*). pp. 195-96.
40. *Ibid.*, p. 196; see also *https://www.bbc.com/news/world-africa-13315719.*
41. *Ibid.*, p. 197.
42. *Ibid.*
43. *https://www.bbc.com/news/world-africa-13315719.*
44. *Ibid.*
45. ILIRIA International Review, Vol. 6, No. 1, 201, (Islam Islami, *Political History of Modern Egypt*). p. 200.
46. *Ibid,* pp. 201-203
47. *Ibid,* p. 203; see also *https://www.britannica.com/place/Egypt/The-Mubarak-regime.*
48. *Ibid.*
49. *https://www.bbc.com/news/world-africa-13315719.*
50. *https://www.telesurenglish.net/analysis/Libya-Before-and-After-Muammar-Gaddafi-20200115-0011.html.*
51. St. John, Ronald Bruce (2006). *Historical Dictionary of Libya*, Oxford, p. xxxii.
52. *Ibid,* p. xxxiii.
53. *Ibid,* p. xxxiv.
54. [The United Nations Development Programme (2010) considered Libya a high-development country in the Middle East and North Africa. This translated status meant a literacy rate of 88.4 per cent, a life expectancy of 74.5 years, gender equality, among several other positive indicators. *https://www.telesurenglish.net/analysis/Libya-Before-and-After-Muammar-Gaddafi-20200115-0011.html.*]
55. *https://eurasiantimes.com/how-a-progressive-libya-under-gaddafi-was-destroyed-the-us-uk-and-france-expert-review/*
56. *https://www.britannica.com/biography/Muammar-al-Qaddafi.*
57. *Ibid.*
58. *Ibid.* (As rebel forces solidified their hold on Tripoli, they intensified their efforts to track down Gaddafi, offering a $1.7 million reward for killing or

capturing him. Gaddafi was killed in Sirte on October 20 as rebel forces took control of the city, one of the last remaining loyalist strongholds.)

59. *https://eurasiantimes.com/how-a-progressive-libya-under-gaddafi-was-destroyed-the-us-uk-and-france-expert-review/*
60. Nuri Eren (1963). *Turkey Today and Tomorrow: An Experiment in Westernization*, New York, p. 40.
61. *Ibid.*, pp. 22-23.
62. *Ibid.*
63. *Ibid.*, pp. 27 & 40.
64. Hashemi, Nader (2009). *Islam, Secularism, and Liberal Democracy*, New York, p. 153. (In 1994, religious icon Erdogan was elected mayor of Istanbul on the ticket of the Welfare Party. The election of the first-ever Islamist to the mayoralty shook the secularist establishment, but Erdogan proved to be a competent and canny manager. He yielded to protests against the building of a mosque in the city's central square but banned the sale of alcoholic beverages in city-owned cafes. In 1998, he was convicted for inciting religious hatred after reciting a poem that compared mosques to barracks, minarets to bayonets, and the faithful to an army. Sentenced to 10 months in prison, Erdogan resigned as mayor. *https://www.britannica.com/biography/Recep-Tayyip-Erdogan*)
65. *https://www.britannica.com/biography/Recep-Tayyip-Erdogan.*
66. Hashemi, Nader (2009). *Islam, Secularism, and Liberal Democracy*, New York, pp. 155-56. In Turkey, there was a mature civil society which consolidated liberal democracy. Even the Islamic-oriented society followed the liberal teachings of Jalaluddin Rumi (1207-1273), Yunus Emre (1238-1320), and the writings of Said Nursi (1878-1960). Even the most radical Islamic movement of Fethullah Gulen believed that religious consciousness should be formed and perpetuated through engaging in social practices and institutions. Their interpretation of Islam was decidedly tolerant, pluralistic, modern, and compassionate. It strives to promote an understanding of faith that is modern, democratic, and secular.
67. Berk Esena and Sebnem Gumuscub, *Rising competitive authoritarianism in Turkey, Third World Quarterly*, 2016, Vol. 37, No. 9, pp. 1581-86.
68. *https://www.britannica.com/biography/Recep-Tayyip-Erdogan.*
69. *https://www.britannica.com/event/Syrian-Civil-War.*
70. *https://www.bbc.com/news/world-middle-east-35806229.*
71. *Ibid.*
72. *Ibid.*
73. *Ibid.*
74. *Islam and Modern Science* by Seyyid Hossein Nasr, *https://www.academia.edu/2521834/Islam_and_Modern_Science_1982.*
75. *https://www.bbc.com/culture/article/20191014-abdus-salam-the-muslim-science-genius-forgotten-by-history.*
76. Muhammad Abdus Salam (1994). *Renaissance of Sciences in Islamic Countries*, Singapore. p. ix.

77. *https://www.britannica.com/place/Palestine/Roman-Palestine.*
78. Al-Djazairi, *op. cit.*, p. 1.
79. *Ibid.*
80. *Ibid.*
81. *https://www.britannica.com/place/Palestine/Roman-Palestine.*
82. *Ibid.* (In July 1922, the Council of the League of Nations approved the mandate instrument for Palestine, including its preamble incorporating the Balfour Declaration and stressing the Jewish historical connection with Palestine. A fresh dispute in Jerusalem concerning religious practices at the Western Wall, sacred to Jews as the only remnant of the Second Temple of Jerusalem and to Muslims as the site of the Dome of the Rock, flared up into serious communal clashes.)
83. *Ibid.*
84. *Ibid.*
85. *Ibid.*
86. *Ibid.*
87. *Ibid.*

Bibliography

Websites and Journals

1. *American Scientist*, 90(4), (July/August 2002).
2. *Diroyah*: Jurnal Studi Ilmu Hadis 5, 1 (2020).
3. *ILIRIA International Review*, Vol. 6, No. 1, 201.
4. *Impact of Science on Society*, (1976), 26(3).
5. International Education Journal, 2006, 7(4).
6. *J Appl Physiol* (1985). 2008 December; 105(6).
7. Revelation and Science, (*Inter-Disciplinary Journal of Intellectual Revival*, International Islamic University, Pahang, Malaysia), 2013, Vol. 3, No. 1.
8. *The FASEB Journal*, Vol. 20, August 2006.
9. *Third World Quarterly*, 2016, Vol. 37, No. 9.
10. Washington Post, December 30, 2001.
11. *Frontier Aging Neuroscience*, 22 February 2022, *https://doi.org/10.3389/fnagi.2022.813531.*
12. *https//en.m.wikipedia.org/wiki/Buyid_dynasty.*
13. *https://eurasiantimes.com/how-a-progressive-libya-under-gaddafi-was-destroyed-the-us-uk-and-france-expert-review/*
14. *https://inews.co.uk/news/world/russia-afghanistan-why-invade-soviet-union-invasion-1979-timeline-what-happened-1156206.*
15. *https://iranicaonline.org/articles/sanjar.*
16. *https://muslimheritage.com/al-mahassin-al-urdi-al-lubudi-al-halabi/*
17. *https://www.academia.edu/2521834/Islam_and_Modern_Science_1982.*
18. *https://www.academia.edu/37479560/CHAPTER_III_BIOGRAPHY_OF_SYED_MUHAMMAD_NAQUIB_AL_ATTAS_AND_HIS_THOUGHT_TOWARD_ISLAMIZATION_OF_KNOWLEDGE.*

19. *https://www.bbc.com/culture/article/20191014-abdus-salam-the-muslim-science-genius-forgotten-by-history.*
20. *https://www.bbc.com/news/world-africa-13315719.*
21. *https://www.bbc.com/news/world-middle-east-35806229.*
22. *https://www.britannica.com/biography/Muammar-al-Qaddafi.*
23. *https://www.britannica.com/biography/Recep-Tayyip-Erdogan.*
24. *https://www.britannica.com/event/Syrian-Civil-War.*
25. *https://www.britannica.com/place/Egypt/The-Mubarak-regime.*
26. *https://www.britannica.com/place/Iran.*
27. *https://www.britannica.com/place/Iraq/Iraq-under-Saddam-Hussein.*
28. *https://www.britannica.com/place/Palestine/Roman-Palestine.*
29. *https://www.britannica.com/topic/Tatar.*
30. *https://www.forbés.com/sites/dominicdudley/2018/02/26/ten-strongest-military-forces-middle-east/?sh=2d7c597f16a2.*
31. *https://www.researchgate.net/publication/331272985_Halabi.*
32. *https://www.telesurenglish.net/analysis/Libya-Before-and-After-Muammar-Gaddafi-20200115-0011.html.*
33. *https://www.timesnownews.com/viral/life-might-actually-flash-before-our-eyes-as-we-die-study-article-89822967 Al-Quran.* .

Books

1. Ahmad, Imam Ibrahim (1967 Eng. Trans.). *Kitab Tahdid al-Amakin of Al-Biruni*, Beirut.
2. Ahmad, Qeyamuddin (1988). *India by Al-Biruni*, New Delhi.
3. *Al-Quran.*
4. Ameer Ali, Syed (1922). *Spirit of Islam: A History of the Evolution and Ideals of Islam with a Life of the Prophet*, London.
5. Ar-Rafai, Shaikh Mohammad Nasib (1998). *Tafsir Ibn Kathir*, (Part-I), London.
6. Beg, Muhammad Abdul Jabbar (ND). *The Origins of Islamic Science*, Leicestershire.
7. Beshore, G. (1998). *Science in Early Islamic Culture*. New York.
8. Bhargava, K.D. (1961). *A Survey of Islamic Culture and Institutions*, Allahabad.
9. Chaney, Eric (2008). *Tolerance, Religious Competition and the Rise and Fall of Muslim Science*. Harvard University.
10. Coughlin, Con (2005). *Saddam: His Rise and Fall*, New York.
11. Dahlan, Winai (2015). *Muhammad: The World's Great Scientist*, Bangkok.

12. Dorsey Armstrong (2014). *Great Minds of the Mediaeval World*, Virginia.
13. El-Fandy, M.G. (1992). *Islam and Science*, Vol. 6, Cairo.
14. Fisk, Robert (2005). *The Great War for Civilisation: The Conquest of the Middle East.* London.
15. Hamid Dabashi (2007). *Iran: A People Interrupted*, New York.
16. Hashemi, Nader (2009). *Islam, Secularism, and Liberal Democracy*, New York.
17. Hoodbhoy, Pervez (1991). *Islam and Science: Religious Orthodoxy and the Battle for Rationality*, London.
18. Hunt Janin (2005). *The Pursuit of Learning in the Islamic World 610-2003*, McFarland & Co Inc., USA.
19. Husaini, S. Waqar Ahmed (1980). *Islamic Environmental Systems Engineering*, London.
20. Iqbal, Muzaffar (2009). *The Making of Islamic Science*, Kuala Lumpur.
21. Kabir, Humayun (1955). *Science, Democracy and Islam*, London.
22. Kamali, Mohammad Hashim (2016). *Islamic Perspectives on Science and Technology*, Singapore.
23. Keith L. Moore (1982). *The Developing Human: Clinically Oriented Embryology*, Philadelphia.
24. M. Rodinson (1966). *Islam and Capitalism*, Paris.
25. M.A. Cook (1970). *Studies in the Economic History of the Middle East*, Oxford.
26. Masood, Ehsan (2009). *Science & Islam: A History of Icon Books*, London.
27. Maurice Bucaille (2000). *The Bible, the Quran, and Science: The Holy Scriptures Examined in the Light of Modern Knowledge*, New Delhi.
28. Miri Shefer-Mossensohn (2015). *Science among the Ottomans: The Cultural Creation and Exchange of Knowledge*, Austin.
29. Miri Shefer-Mossensohn (2015). *Science among the Ottomans: The Cultural Creation and Exchange of Knowledge*, Austin.
30. Mishkat al-Masabih.
31. Muhammad Abdus Salam (1994). *Renaissance of Sciences in Islamic Countries*, Singapore.
32. Nasr, Scyyed Hossein (1976). *Islamic Science: An Illustrated Study*, London.
33. Nasr, Seyyed Hossein (2001). *Science and Civilization in Islam*, Chicago.

34. Nasr, Seyyed Hossein (2010). *In Search of the Sacred: A Conversation with Seyyed Hossein Nasr on his Life and Thought*, Santa Barbara, California.
35. Nuri Eren (1963). *Turkey Today and Tomorrow: An Experiment in Westernization*, New York. Hashemi, Nader (2009). *Islam, Secularism, and Liberal Democracy*, New York.
36. Roberts, Paul (2005). *The End of Oil: On the Edge of a Perilous New World.* Boston.
37. S.E. Al-Djazairi (2006). *The Golden Age and Decline of Islamic Civilisation*, Manchester.
38. *Sahih Al-Bukhari.*
39. *Sahih al-Jami.*
40. *Sahih Muslim.*
41. *Sahih Musnad Ahòmad.*
42. Salim T.S. Al-Hassani (3rd Edition, 2012). *1001 Inventions: The Enduring Legacy of Muslim Civilization*, Washington DC.
43. St. John, Ronald Bruce (2006). *Historical Dictionary of Libya*, Oxford.
44. *Sunan Abu Daud.*
45. *Sunan al-Tirmidhi.*
46. *Sunan an-Nasai.*
47. *Sunan Ibn Majah.*
48. Tripp, Charles (2010), *A History of Iraq*, Cambridge.
49. Turner, R.H. (1995). *Science in Mediaeval Islam: An Illustrated Introduction. Austin.*
50. W. Durant (1950). *The Age of Faith,* New York, p. 208; see also A. Pacey (1990). *Technology in World Civilization: A Thousand Year History,* Cambridge.
51. Walbridge, John (2011). *God and Logic in Islam: The Caliphate of Reason,* New York.
52. Wan Hazmy C. H., Zainurashid Z., Hussaini R. (2009). *Biography: Muslim Scholars and Scientists,* N. Sembilan.

Index